Solved Problems in Statistical Inference

Charles Ashbacher
5530 Kacena Ave
Marion, IA 52302 USA

cashbacher@yahoo.com

Artwork

Caytie Ribble

ISBN: 978-1515215622

Contents

Introduction

This book grew out of the most common request that I have received in my years of teaching basic statistics. "Give us more problems on statisticial inference." Generally speaking, my class in statistics is split into two sections: descriptive statistics plus basic probability and inferential statistics or hypothesis testing.

The first section of the class where descriptive statistics and probability are covered tend to be easy for the students to understand. I do not require the memorization of expressions, so for the most part the students simply follow the formula and perform the computation.

However, when the focus shifts to performing inferences, some students seem to hit the wall. Where they had been moving through the material with complete understanding, they are now puzzled by the new form of thinking. Most get through it, but not without a bit of struggling. Extra problems with solutions generally provides them with enough additional assistance to help them get through the material.

There is no attempt to be comprehensive in terms of the types of inference problems that can appear in a basic statistics course. The problems solved here are just those that are covered in my class. I hope that you find them as helpful as my students have.

As always, I welcome comments and feedback to my email address below.

Charles Ashbacher

cashbacher@yahoo.com

Cartoons by Caytie Ribble

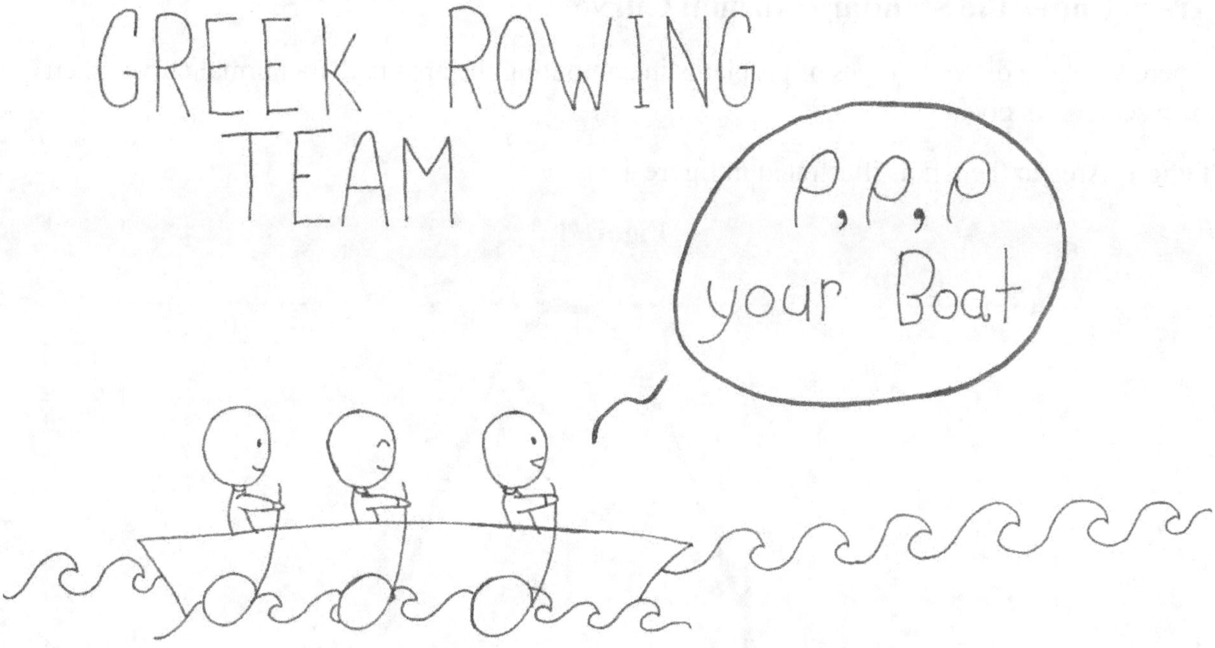
GREEK ROWING TEAM
ρ, ρ, ρ your Boat

Areas Under the Standard Normal Curve

There are three different types of problems in computing the area under a normal curve when the z score(s) is given.

Type 1: Area to the left as illustrated in figure 1.

Figure 1

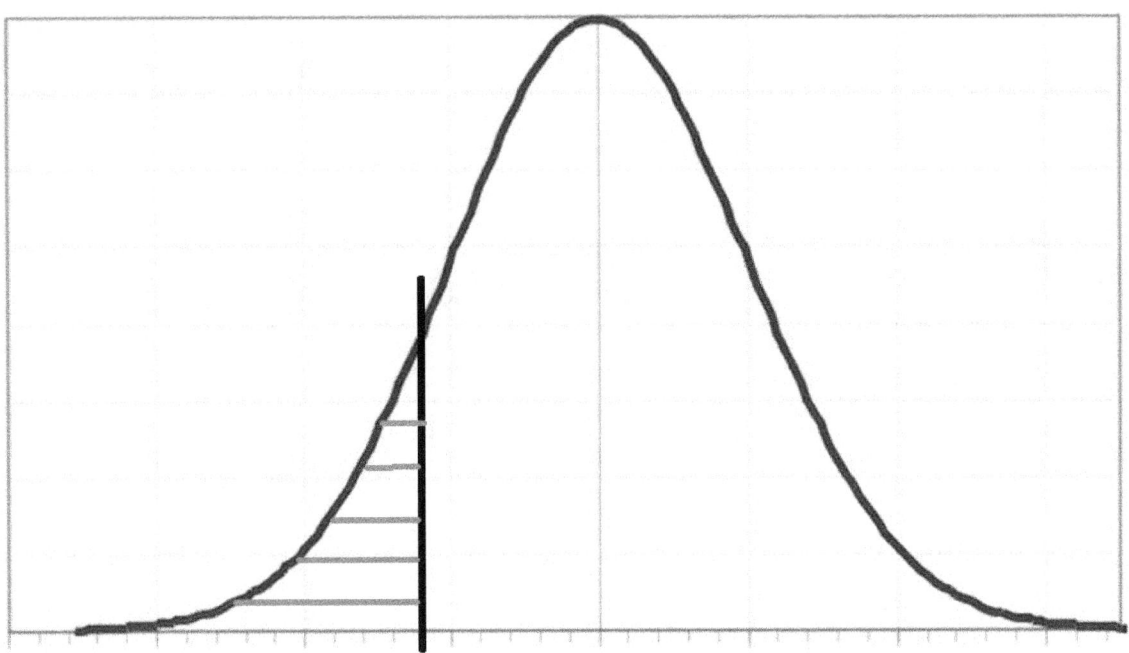

In this case, you find the z value in Table A and read the area directly from the chart,

1. Find the area of z < -1.67.

2. Find the area of z < 1.19.

Type 2: Area to the right as illustrated in figure 2.

In this problem we take advantage of the fact that geometric areas can be added and that the sum of the area under the curve is 1.0000. We look up the area to the left of the position in the table and then subtract it from 1.0000 to get the area to the right.

3. Find the area to the right of z = -2.11.

4. Find the area to the right of z = 1.55.

Figure 2

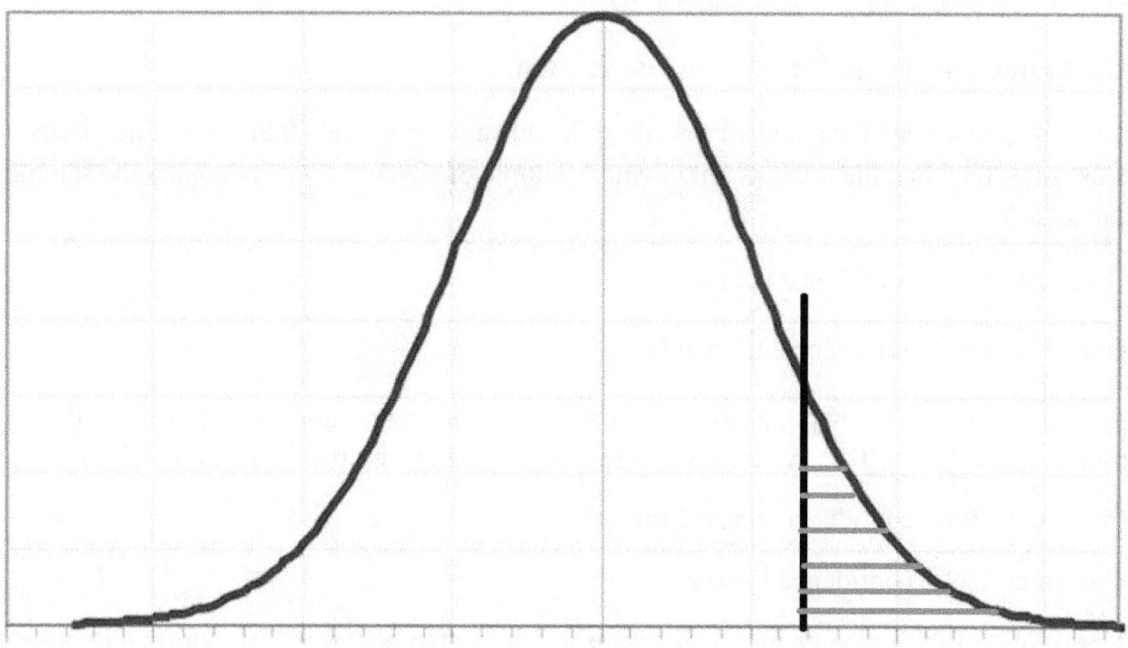

Type 3: The area between two positions, as illustrated in figure 3.

Figure 3

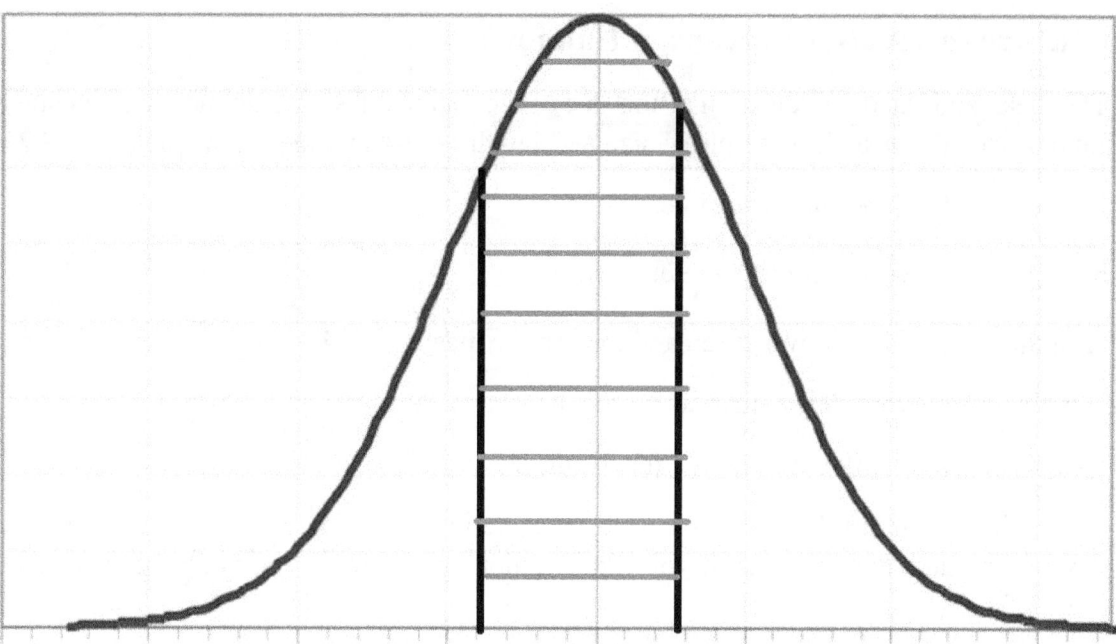

In this case, the geometric principle that areas can be subtracted is used. The largest and the smallest areas are determined from the table and then the smallest subtracted from the largest.

5. Find the area between z = -1.48 and z = 1.22.

6. Find the area between z = 1.11 and z = 3.12.

Confidence Intervals for Means, σ Known

7. It is well known that the standard deviation for the amount of soda that a machine places in bottles is σ = 0.65 and the mean is unknown. A sample of size n = 49 is taken and the sample mean is 24.3.

a) Construct a 95% confidence interval for μ

b) Construct a 99% confidence interval for μ

8. A machine places a very popular cereal in boxes and the way it does so can be described by a normal curve with σ = 0.25. A sample of size n = 36 is taken and the sample mean is 20.3.

a) Construct a 95% confidence interval for μ.

b) Construct a 99% confidence interval for μ.

9. The weight of third graders in a large urban school district is defined by a normal curve with σ = 3.6. A sample of size n = 25 is taken and the sample mean is 56.3.

a) Construct a 95% confidence interval for μ.

b) Construct a 99% confidence interval for μ.

Confidence Intervals for Means, σ Unknown

10. The mean and standard deviation of the size of the fish in a lake are unknown. In an attempt to find out what the mean is a sample of size n=25 with s = 1.4 and the sample mean is 18.2 oz.

   a) Construct a 95% confidence interval for μ

   b) Construct a 99% confidence interval for μ

11. A sample of size n = 16 was taken and the sample mean was 24.5 and s = 2.5.

a) Construct a 95% confidence interval for the mean.

b) Construct a 99% confidence interval for the mean.

12. A machine that is designed to place 20 oz of liquid in bottles can be described by a normal curve where μ and σ are unknown. If you take a sample of size n = 30, s = 0.9 and the mean is 18.9:

a) Construct a 95% confidence interval for μ.

b) Construct a 99% confidence interval for μ.

Hypothesis testing of the mean, σ known

13. A nutritionist believes that the mean weight of third graders has gone down from the previously known value of $\mu = 59.5$ pounds. It is known that $\sigma = 4.2$ and a sample of size $n = 36$ is taken and the sample mean is 53.6.

    a) What can be concluded at the $\alpha = 0.05$ level?

    b) What can be concluded at the $\alpha = 0.01$ level?

14. You work where a machine places oatmeal in boxes. It is known that the machine is defined by a normal curve with $\sigma = 1.1$. The mean is supposed to be 24 oz but you believe that the machine is malfunctioning. You take a sample of size $n=36$ and the sample mean is 24.5.

a) What can you conclude at the $\alpha = 0.05$ level?

b) What can you conclude at the $\alpha = 0.01$ level?

15. The manager of a production line where liquid is placed in bottles by a machine believes that the machine is malfunctioning, putting too much in the bottles. If the machine is supposed to put $\mu = 16$ oz. in the bottles with $\sigma = 0.5$ and a sample of size $n = 36$ is taken with the sample mean equal 16.6, what can the manager conclude.

a) At the $\alpha = 0.05$ level of confidence?

b) At the $\alpha = 0.01$ level of confidence?

16. A machine that puts cereal in boxes is supposed to put in an average of 20 oz. You believe that the machine is malfunctioning, so you take a sample of size $n = 25$ and get a sample mean of 20.25 oz. If $\sigma = 0.6$:

a) What can you conclude at the $\alpha = 0.05$ level of confidence?

b) What can you conclude at the $\alpha = 0.01$ level of confidence?

Hypothesis testing of the mean, σ unknown

17. A quality control worker believes that a machine that places liquid in bottles is malfunctioning; she believes that it is in fact putting more than the designed 16 oz. in the bottles on average. To test her belief she takes a sample of size $n = 25$, $s = 1.5$ with the sample mean equal to 16.5.

a) What can she conclude at the $\alpha = 0.05$ level of confidence?

b) What can she conclude at the $\alpha = 0.05$ level of confidence?

18. A researcher believes that the average weight of fourth graders has dropped from 84 pounds due to the new lunch menu. She takes a random sample of size n = 16 where the sample mean is 82 and s = 3.6.

a) What can she conclude at the $\alpha = 0.05$ level?

b) What can she conclude at the $\alpha = 0.01$ level?

19. A biologist believes that the increased level of carbon dioxide in the air is causing an increase in the average weight of Morrell mushrooms this spring. It has been well documented that the average weight of the mushrooms last year was 3.2 oz. If a sample of size n = 18 is taken and the sample mean is 3.4 and s = 1.2, what can the biologist conclude at

a) The $\alpha = 0.05$ level?

b) The $\alpha = 0.01$ level?

20. An educational researcher believes that the average height of sunflowers in a garden is different from what the seller of the seeds claimed that they would be. The claim by the seed seller was that they would average six feet (72 inches) in height. She took a random sample of size 20 and the mean was 64 inches with s = 3.2.

a) What can the researcher conclude at the $\alpha = 0.05$ level of confidence?

b) What can the researcher conclude at the $\alpha = 0.01$ level of confidence?

Confidence intervals for the difference between two means, σ_1 and σ_2 known

21. Two different timing mechanisms are being used to measure the results of rowing races and there is a bit of variance it their performance. The interest is in determining the difference in the average performance of the two methods. If $\sigma_1 = 0.3$ and $\sigma_2 = 0.25$ and the sample sizes are $n_1 = 25$ and $n_2 = 36$ and the mean of the first is 88.5 and the mean of the second is 88.4

a) Construct a 95% confidence interval for μ.

b) Construct a 99% confidence interval for μ.

22. Two algebra classes are being taught, one (class number 1) using a traditional method and the other (class number 2) using a new, experimental method. Each class is given the same 200 point exam at the end and the mean score for the first class is 189 and the mean score for the second class is 194. If $n_1 = 24$ and $n_2 = 28$ and $\sigma_1 = 2.5$, $\sigma_2 = 2.1$, construct the following:

a) A 95% confidence interval for the difference of the two means.

b) A 99% confidence interval for the difference of the two means.

23. It is well known that a certain type of machine puts cereal in boxes with $\sigma = 0.85$. Twenty-three boxes from one machine are tested and the average amount in them was 16.1 oz. and 32 boxes from another machine were tested and the average amount in them was 15.9 oz.

a) Construct a 95% confidence interval for the difference in the two means.

b) Construct a 99% confidence interval for the difference in the two means.

Confidence intervals for the difference between two means, σ_1 and σ_2 unknown

24. A 200 point exam is given in two different classes where the method of instruction differed. The data for the two classes is

$mean_1 = 185.4$, $n_1 = 18$, $s_1 = 3.7$

$mean_2 = 190.4$, $n_2 = 21$, $s_2 = 4.1$.

a) Construct a 95% confidence interval for the difference of the two means.

b) Construct a 99% confidence interval for the difference of the two means.

25. Two different pain relievers are given to two distinct groups and the level of pain they felt after consumption was recorded on a scale of one to one hundred. The data for the two groups is

$mean_1 = 63.5$, $n_1 = 14$, $s_1 = 2.4$

$mean_2 = 58.9$, $n_2 = 16$, $s_2 = 3.1$.

a) Construct a 95% confidence interval for the difference of the two means.

b) Construct a 99% confidence interval for the difference of the two means.

26. A study is done to determine the strength of a new sealant for water pipes. Twelve pipes are connected using sealant 1 and the strength of the connections is measured. The mean is 82.4 with $s_1 = 3.3$. Fourteen pipes are connected using sealant 2 and the mean is 84.8 with $s_2 = 4.1$.

a) Construct a 95% confidence interval for the difference of the two means.

b) Construct a 99% confidence interval for the difference of the two means.

Testing the difference between two means, σ_1 and σ_2 known

27. The machines that put soda into bottles (goal of $\mu = 20$) at a company are consistent performers with a standard deviation of the amount the put in the bottles at $\sigma = 0.4$. A member of the quality control staff believes that two machines are putting different amounts in bottles. To test this she takes two random sample of size $n = 36$ from each of the machines. If the mean of the first sample is 16.1 and the mean of the second sample is 15.9:

a) What can she conclude at the $\alpha = 0.05$ level of confidence?

b) What can she conclude at the $\alpha = 0.01$ level of confidence?

28. After years of study it is known that the standard deviation of the average weight of letters sent from businesses is 0.18 and that the standard deviation of the average weight of letters sent from private residences is 1.12. A sample of size 36 is taken of letters sent from businesses and the mean was 3.25 and a sample of size 25 is taken from private residences and the mean was 3.11. A researcher believes that the average weight of letters from businesses is greater than the average weight of letters from private businesses.

a) What can he conclude at the $\alpha = 0.05$ level of confidence?

b) What can he conclude at the $\alpha = 0.01$ level of confidence?

29. Two different cumulative exams in algebra have been given for many years and the standard deviations of both are well known. For the first exam $\sigma_1 = 2.65$ and for the second exam $\sigma_2 = 3.15$. An educational researcher believes that the scores on the second exam are better so she takes a random sample of size 28 from the first group and a random sample of size 30 from the second. The mean from the first group is 92.4 and the mean from the second is 94.2.

a) What can she conclude at the $\alpha = 0.05$ level of confidence?

b) What can she conclude at the $\alpha = 0.01$ level of confidence?

Testing the difference between two means, σ_1 and σ_2 unknown

30. Two different drugs designed to reduce the level of pain are to be tested against each other, specifically the belief that they are not of the same effectiveness in reducing the level of pain. A total of 46 people were split into two groups and in a double-blind experiment, 20 were given A and 26 given B. All people were questioned using a simple pain indicator scale of one through 20 (higher means more pain) and the the mean pain scale for group A was 15.4 while the mean pain scale for group B was 14.2. If the standard deviation for the A group was 2.1 and the standard deviation for the B group was 2.8, what can be concluded:

a) At the $\alpha = 0.05$ level?

b) At the $\alpha = 0.01$ level?

31. A teacher believes that she has developed an educational strategy that is far superior to what has been used. The new technique is applied to group 1 and the old technique is used with group 2. The data is

$Mean_1 = 86.5$, $n_1 = 21$, $s_1 = 3.6$

$Mean_2 = 80.2$, $n_2 = 19$, $s_2 = 2.9$.

a) What can she conclude at the $\alpha = 0.05$ level?

b) What can she conclude at the $\alpha = 0.01$ level?

32. A biologist believes that the average weight of frogs in the pond next to a chemical plant is less than that of a pond some distance away. There is no exchange of frogs between the two ponds and to test his hypothesis, the biologist takes samples from each pond. The chemical plant is near where sample 1 was taken and here is the data.

$\text{Mean}_1 = 8.8$, $n_1 = 12$, $s_1 = 1.9$

$\text{Mean}_2 = 10.4$, $n_2 = 13$, $s_2 = 1.7$

a) What can the biologist conclude at the $\alpha = 0.05$ level?

a) What can the biologist conclude at the $\alpha = 0.05$ level?

Confidence intervals for proportions

33. A news organization did a poll of 300 people and 56 expressed a favorable view of a loudmouth politician.

a) Construct a 95% confidence interval for the true proportion of people that have a favorable view of the politican.

a) Construct a 99% confidence interval for the true proportion of people that have a favorable view of the politican.

34. The negotiations for a major treaty between nations have recently been concluded. A public opinion survey was taken and 67% of the 400 people surveyed expressed opposition to the terms of the treaty.

a) Construct a 95% confidence interval for the true proportion.

b) Construct a 99% confidence interval for the true proportion.

35. A polling company conducted a survey of 500 people and only 100 were in favor of the posal service eliminating Saturday delivery.

a) Construct a 95% confidence interval for the true proportion.

b) Construct a 99% confidence interval for the true proportion.

Hypothesis testing with proportions

36. A news media outlet claimed that 55% of the people were in favor of a particular candidate for political office. However, a political science professor believes that this is incorrect and to test it she conducts a random survey of 400 people and 200 of them are in favor of the candidate.

a) What can she conclude at the $\alpha = 0.05$ level?

b) What can she conclude at the α = 0.01 level?

37. It has been a textbook "fact" for decades that the percentage of people that engage in a specific practice is 0.12. A sociologist believes that this is in fact too low. She takes a survey of 1,000 randomly selected people and 210 state that they engage in the practice.

a) What can she conclude at the α = 0.05 level?

b) What can she conclude at the α = 0.01 level?

38. It has been reported in the sociology literature that 55% of people believe that they are middle class. However, a sociology professor believes that this number is too high. To test his belief he surveys 660 people and 320 respond that they believe they are in the middle class.

a) What can she conclude at the α = 0.05 level?

b) What can she conclude at the α = 0.01 level?

Hypothesis testing with correlation coefficients

39. A company claims that their method of preparation for the ACT exam is superior and there is a direct correlation between the hours of instruction and the increase in scores. Eight students are tested and the data is summarized in the following table

Hours of instruction	Score on test
2	25.6
2.5	26
3.0	26.1
3.5	26
3.5	26.4
4.0	26.9
4.5	27.9
5.0	27.3

a) What can be concluded at the α = 0.05 level of confidence?

b) Construct the linear regression line

c) Use the regression line to predict the score if a student has 6 hours of instruction.

40. A social researcher believes that the average number of text messages sent per day has a negative effect on academic performance. To test this she examines a class of 8 students and obtains the following data

Average number of text messages	Score on exam
4.4	90
3.6	92
5.6	88
5.2	88
6.9	86
7.8	84
7.2	82
8.4	83

a) What can she conclude regarding her belief at the $\alpha = 0.05$ level?

b) Determine the equation of the regression line

c) Use the regression equation to predict the test score of someone that sends an average of 10 text messages a day.

41. A nutritionist believes that that the number of hours that a child plays video games leads to increased weight. To test this she gathers the following data for children aged 12

Hours per week playing video games	Weight
2.5	78.4
2.2	77
2.0	73
2.8	79
3.4	85
3.1	87
3.0	86
2.9	79

a) What can she conclude at the $\alpha = 0.01$ level?

b) Determine the linear regression line

c) Use the linear regression line to estimate the weight of a child that plays video games 6 hours a week.

Solutions

Area under standard normal curve

1. Reading the value directly from table A, the area is .0475.

2. Reading the value directly from table A, the area is .8830.

3. From table A, the area to the left of -2.11 is .0174. Subtracting this from 1.0000, the area to the right is 0.9826.

4. From table A, the area to the left of 1.55 is .9394. Subtracting this from 1.0000, the area to the right is 0.0606.

5. From table A the area to the left of -1.48 is .0694 and the area to the left of 1.22 is .8888. Therefore, the area in between is .8888 - .0694 = 0.8194.

6. From table A the area to the left of 1.11 is .8665 and the area to the left of 3.12 is .9991. Therefore, the area in between is .9991 - .8665 = .1326.

Confidence intervals where σ is known

7. In this problem we know the value of σ so we will use a z score in setting the limits.

a) For a 95% confidence interval, we use $z_{\alpha/2} = 1.96$, so the interval is

$$24.3 - 1.96*(0.65/\sqrt{49}) \leq \mu \leq 24.3 + 1.96*(0.65/\sqrt{49}).$$

Computing, the values are

$$24.3 - 0.18 \leq \mu \leq 24.3 + 0.18 \text{ or } 24.12 \leq \mu \leq 24.48.$$

b) For a 99% confidence interval we use $z_{\alpha/2} = 2.58$, so the interval is

$$24.3 - 2.58*(0.65/\sqrt{49}) \leq \mu \leq 24.3 + 2.58*(0.65/\sqrt{49}).$$

Computing, the values are

$$24.3 - 0.24 \leq \mu \leq 24.3 + 0.24 \text{ or } 24.06 \leq \mu \leq 24.54.$$

8. In this problem, we know the value of σ so we will use a z score in setting the limits.

a) For a 95% confidence interval, we use $z_{\alpha/2} = 1.96$, so the interval is

$$20.3 - 1.96*(0.25/\sqrt{36}) \leq \mu \leq 20.3 + 1.96*(0.25/\sqrt{36}).$$

Computing, the values are

$$20.3 - 0.08 \leq \mu \leq 20.3 + 0.08 \text{ or } 20.22 \leq \mu \leq 20.38.$$

b) For a 99% confidence interval we use $z_{\alpha/2} = 2.58$, so the interval is

$$20.3 - 2.58*(0.25/\sqrt{36}) \leq \mu \leq 20.3 + 2.58*(0.25/\sqrt{36}).$$

Computing, the values are

$20.3 - 0.11 \le \mu \le 20.3 + 0.11$ or $20.19 \le \mu \le 20.41$.

9. In this problem, we know the value of σ so we will use a z score in setting the limits.

a) For a 95% confidence interval, we use $z_{\alpha/2} = 1.96$, so the interval is

$56.3 - 1.96*(3.6/\sqrt{25}) \le \mu \le 56.3 + 1.96*(3.6/\sqrt{25})$.

Computing, the values are

$56.3 - 1.41 \le \mu \le 56.3 + 1.41$ or $54.89 \le \mu \le 57.71$.

b) For a 99% confidence interval we use $z_{\alpha/2} = 2.58$, so the interval is

$56.3 - 2.58*(3.6/\sqrt{25}) \le \mu \le 56.3 + 2.58*(3.6/\sqrt{25})$.

Computing, the values are

$56.3 - 1.86 \le \mu \le 56.3 + 1.86$ or $54.44 \le \mu \le 58.16$.

Confidence intervals where σ is not known

10. a) With d. f. = 24, $t_{\alpha/2} = 2.064$, so the interval is

$18.2 - 2.064*(1.4/\sqrt{25}) \le \mu \le 18.2 + 2.064*(1.4/\sqrt{25})$.

Computing, the values are

$18.2 - 0.41 \le \mu \le 18.2 + 0.41$ or $17.79 \le \mu \le 18.61$.

b) With d. f. = 24, $t_{\alpha/2} = 2.797$, so the interval is

$18.2 - 2.797*(1.4/\sqrt{25}) \le \mu \le 18.2 + 2.797*(1.4/\sqrt{25})$.

Computing, the values are

$18.2 - 0.78 \le \mu \le 18.2 + 0.78$ or $17.42 \le \mu \le 18.98$.

11.a) With d. f. = 15, $t_{\alpha/2} = 2.131$, so the interval is

$24.5 - 2.131*(2.5/\sqrt{16}) \le \mu \le 24.5 + 2.131*(2.5/\sqrt{16})$.

Computing, the values are

$24.5 - 1.33 \le \mu \le 24.5 + 1.33$ or $23.17 \le \mu \le 25.83$.

b) With d. f. = 15, $t_{\alpha/2} = 2.947$, so the interval is

$24.5 - 2.947*(2.5/\sqrt{16}) \le \mu \le 24.5 + 2.947*(2.5/\sqrt{16})$.

Computing, the values are

$18.2 - 1.84 \le \mu \le 18.2 + 1.84$ or $16.36 \le \mu \le 20.04$.

12. a) With d. f. = 29, $t_{\alpha/2}$ = 2.045, so the interval is

$18.9 - 2.045(0.9/\sqrt{30}) \leq \mu \leq 18.9 + 2.045(0.9/\sqrt{30})$.

Computing,

$18.9 - 0.34 \leq \mu \leq 18.9 + 0.34$ or $18.56 \leq \mu \leq 19.24$.

b) With d. f. = 29, $t_{\alpha/2}$ = 2.756, so the interval is

$18.9 - 2.756(0.9/\sqrt{30}) \leq \mu \leq 18.9 + 2.756(0.9/\sqrt{30})$.

Computing,

$18.9 - 0.45 \leq \mu \leq 18.9 + 0.45$ or $18.45 \leq \mu \leq 19.35$.

Hypothesis testing with means where σ is known

13. a)

Step 1: Set H_0 and H_1. H_0: μ = 59.5, H_1: μ < 59.5, since this is the belief.

Step 2: Draw the diagram and set the limit(s). This is seen in figure 4. We use a z-score because σ is known.

Step 3: Compute the test value

$$z = \frac{53.6 - 59.5}{(4.2 / \sqrt{36})} = \frac{-5.9}{0.7} = -8.43.$$

Step 4: Compare and conclude

-8.43 is well within the tail so we reject H_0 and accept H_1.

b) All the parameters are the same except for the limit on the tail, which is now -2.33. Therefore, we once again reject H_0 and accept H_1.

14. a)

Step 1: Set H_0 : μ =24, H_1: $\mu \neq 24$, since the belief is that the mean is not 24.

Step 2: Draw the diagram and establish the limit(s). This is a two-tailed test with z-scores so the situation is summarized in figure 5.

Step 3: Compute the test value

$$z = \frac{24.5 - 24}{(1.1 / \sqrt{36})} = \frac{0.5}{0.18} = 2.78.$$

19

Figure 4

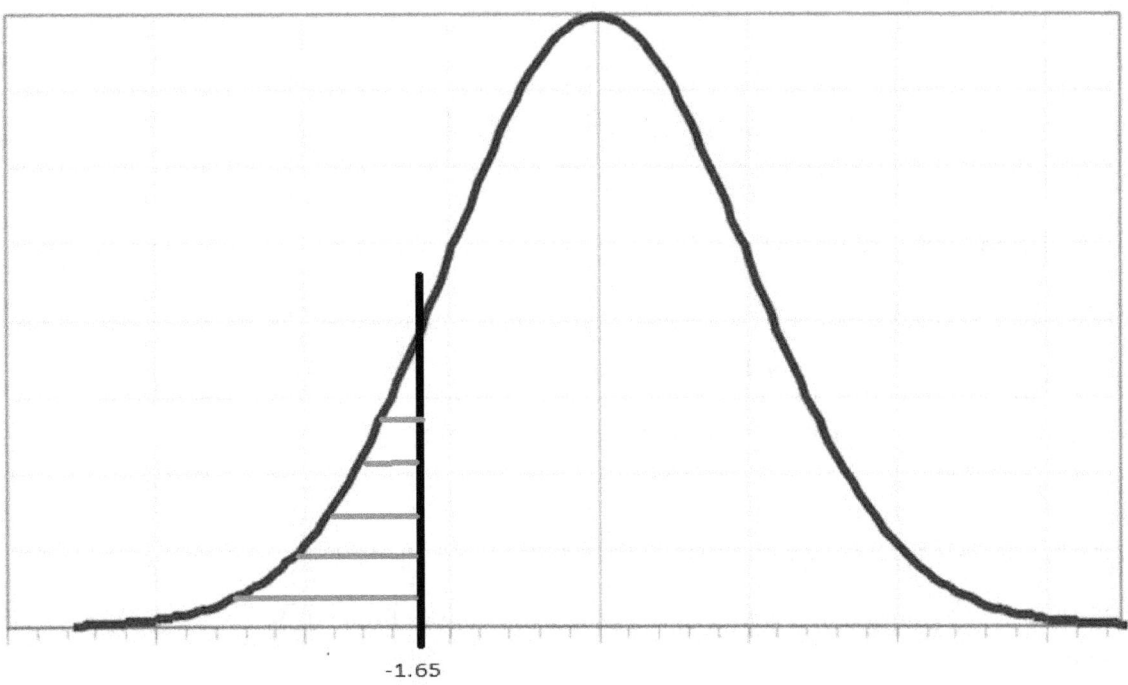

-1.65

Figure 5

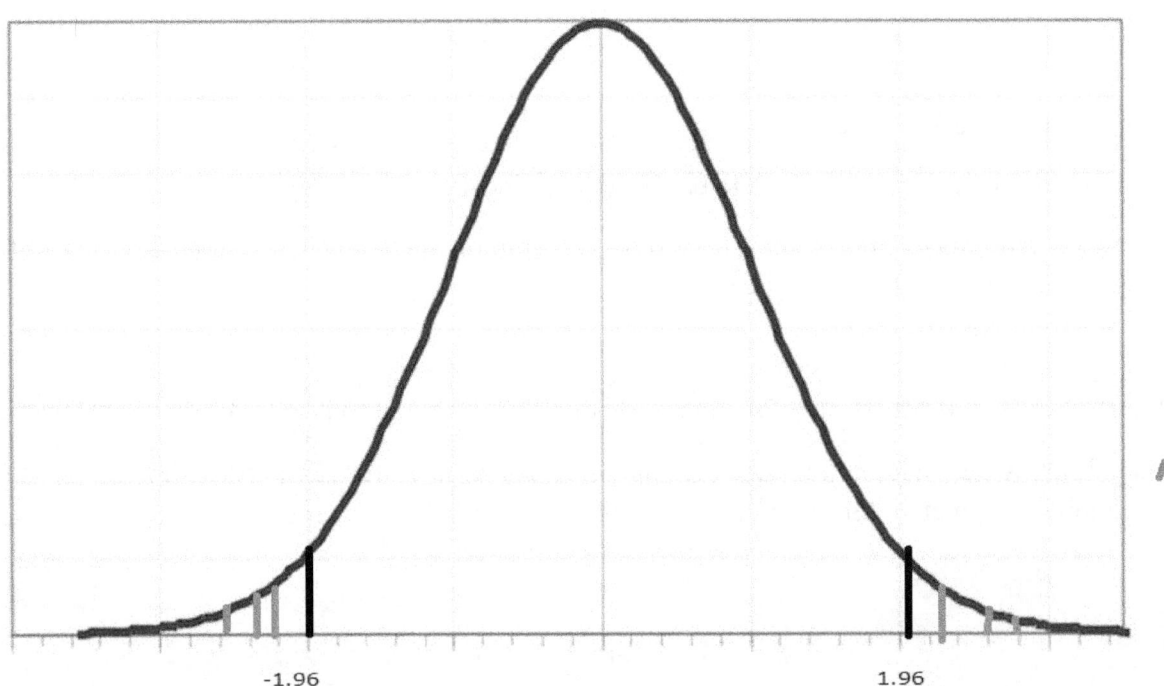

-1.96   1.96

Step 4: Compare and conclude. Since 2.78 is larger than 1.96, we reject H_0 and accept H_1.

b) The only parameter that has changed is the limits on the tails, which is now 2.58. Since 2.78 is larger, we once again reject H_0 and accept H_1.

15. a)

Step 1: H_0 $\mu = 16$, H_1 $\mu > 16$, since this is the belief.

Step 2: This is a one-tailed test on the right, the situation is represented by figure 6.

Figure 6

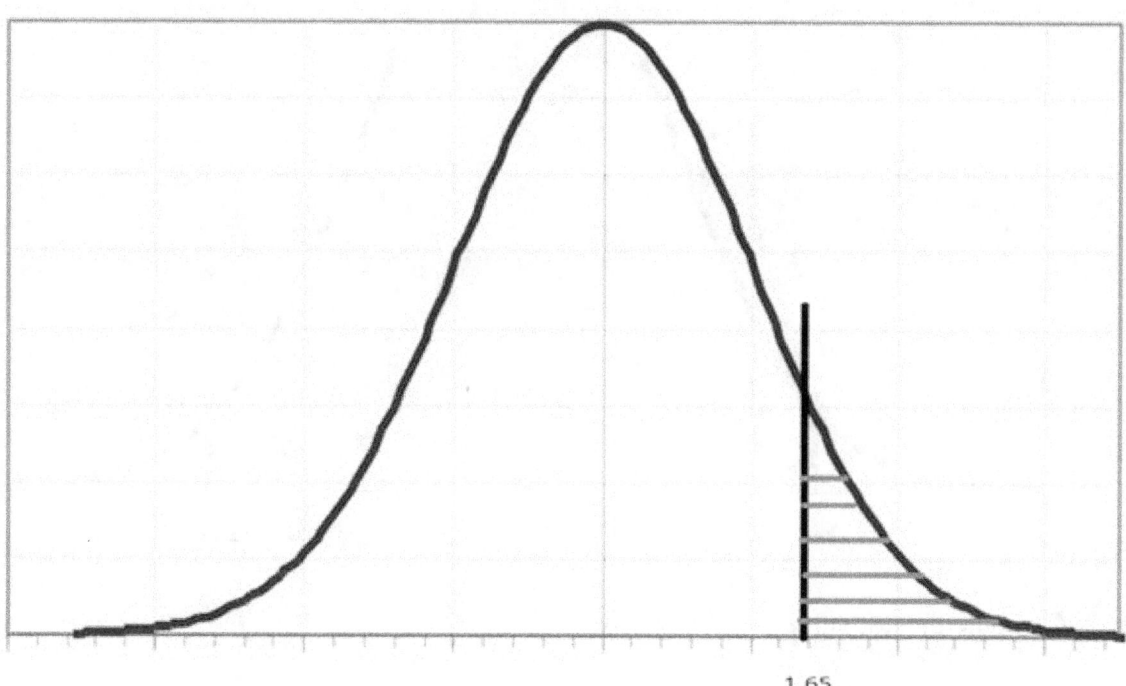

1.65

Step 3: Conpute the test value

$$z = \frac{16.6 - 16}{(0.5/\sqrt{36})} = \frac{0.6}{0.08} = 7.5$$

Step 4: Compare and conclude. In this case, we have landed in a tail, so we reject H_0 and accept H_1.

b) The limit for the $\alpha = 0.01$ case is 2.33 so we also reject H_0 and accept H_1 at this level.

16. Since we know σ we use z scores.

a) Step 1: determine the null and alternative hypotheses

21

H_0: $\mu = 20$, H_1: $\mu \neq 20$, since the belief is that the machine is putting an average other than 20 in the boxes.

Step 2: Draw the diagram and set the limit(s). Since this is a two-tailed test for $\alpha = 0.05$, we use the curve in figure 7.

Figure 7

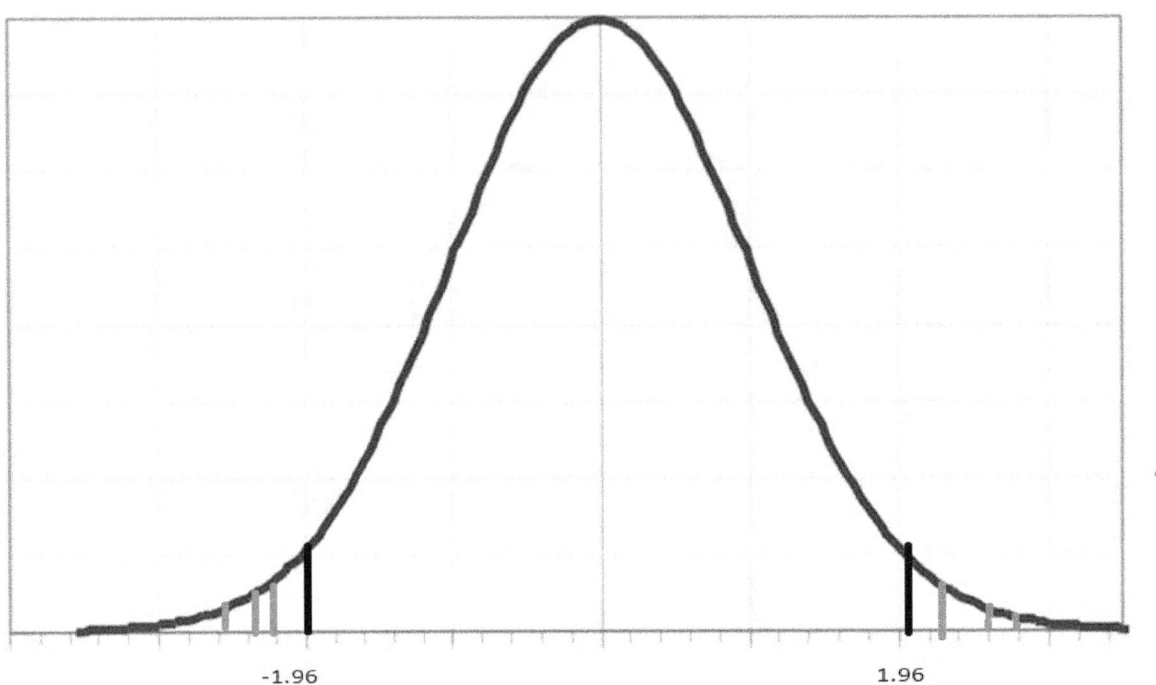

-1.96   1.96

Step 3: Compute the test value

$$z = \frac{20.25 - 20}{(0.6 / \sqrt{25})} = \frac{0.25}{0.12} = 2.08.$$

Step 4: Compare and conclude. The value is greater than 1.96 so we reject H_0 and accept H_1.

b) The only thing different is that the limits for the tails are now -2.58 and 2.58. Therefore, at this level we DO NOT reject H_0.

Hypothesis testing with means where σ is unknown

17. a)

Step 1: Establish H_0: $\mu = 16$, H_1: $\mu \neq 16$.

Step 2: Draw the diagram and establish the limits. This is a two-tailed test using t-scores since σ is unknown. Since n = 25, d.f. = 24 and from table B we have the limits seen in figure 8.

Step 3: Compute the test value

$$t = \frac{16.5 - 16}{(1.5 / \sqrt{25})} = \frac{0.5}{0.3} = 1.67.$$

Step 4: Compare and conclude. Since the test value is not in a tail we CANNOT reject H_0.

Figure 8

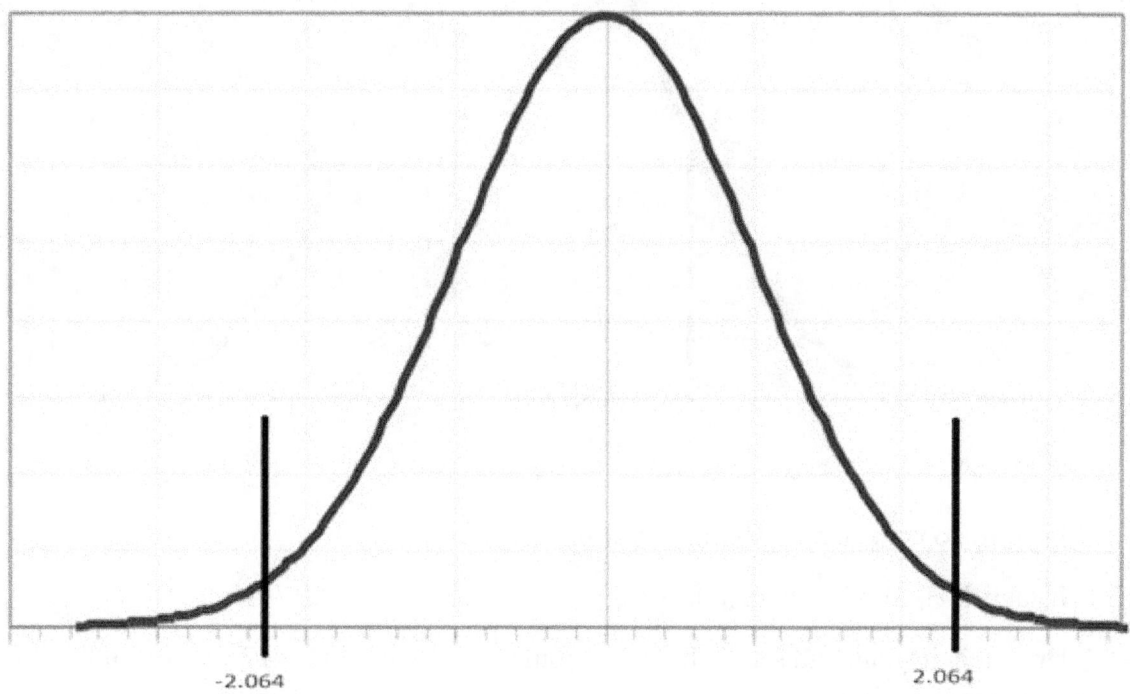

-2.064   2.064

b) Since we could not reject H_0 at the $\alpha = 0.05$ level we CANNOT reject it at the 0.01 level. This is a general rule of hypothesis testing.

18. a)

Step 1: Establish $H_0 = 84$, $H_1 < 84$

Step 2: Draw the diagram and establish the limit(s). From table B the one tail-value for $\alpha = 0.05$ with d. f. = 15 is 1.753, this is represented in figure 9.

Step 3: Compute the test value

$$z = \frac{82 - 84}{(3.6 / \sqrt{16})} = \frac{-2}{0.9} = -2.22$$

23

Step 3: Compare and conclude. Since -2.22 is in the tail, we reject H_0 and accept H_1.

b) From table B the one-tail value for $\alpha = 0.01$ is -2.602, so we do NOT reject H_0 at the $\alpha = 0.01$ level.

Figure 9

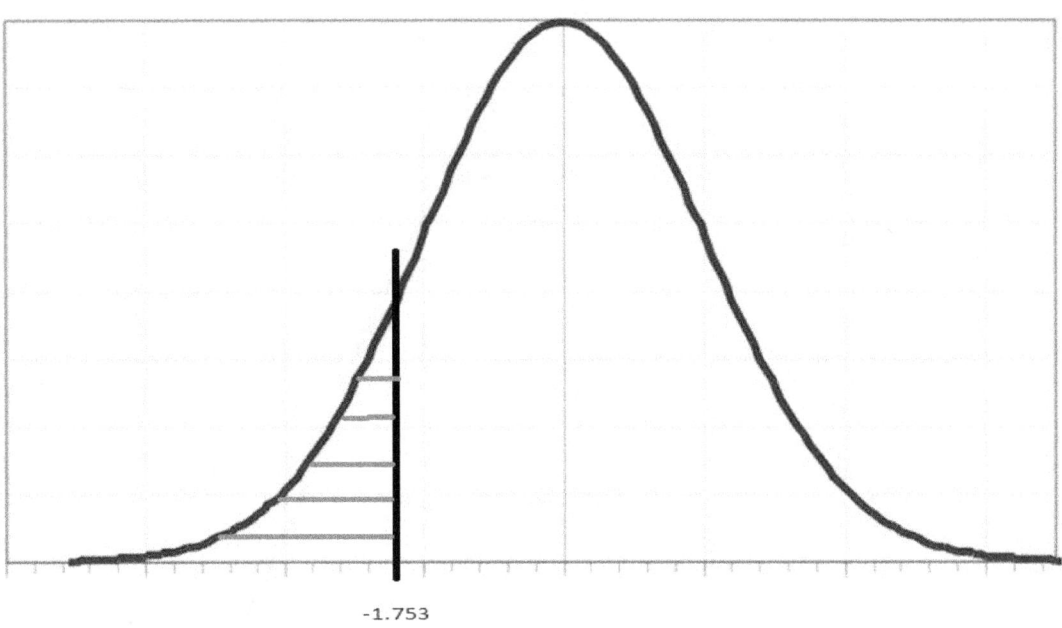

-1.753

19. a) Since we do not know σ, we must use t scores.

Step 1: Establish H_0: $\mu = 3.2$ and H_1: $\mu > 3.2$.

Step 2: Draw the diagram and set the limit(s). From table B with d.f.. = 17 with one tail the value is 1.740, expressed in figure 10.

Step 3: Compute the test value

$$z = \frac{3.4 - 3.2}{(1.2 / \sqrt{18})} = \frac{0.2}{0.28} = 0.71.$$

Step 4: Compare and conclude

Since 0.71 is not in a tail, we do NOT reject H_0.

b) Since we did not reject H_0 at the $\alpha = 0.05$ level, we cannot reject it at the $\alpha = 0.01$ level.

20. Since we do not know σ, we must use t scores.

a)

Step 1: Establish H_0: $\mu = 72$, H_1: $\mu \neq 72$.

Figure 10

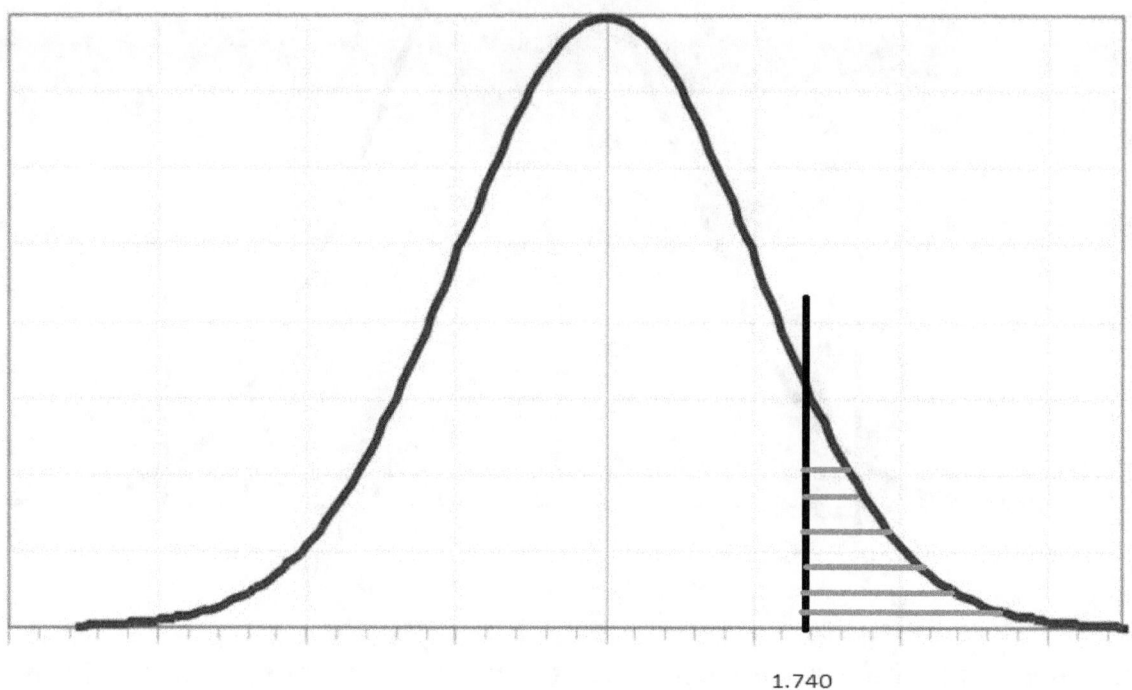

1.740

Step 2: Draw the diagram and set the limits. This is a two-tailed test with d.f. = 19 and the limits derived from table B appear in figure 11.

Step 3: Compute the test value

$$z = \frac{64 - 72}{(3.2 / \sqrt{20})} = \frac{-8}{0.72} = -11.1.$$

Step 4: Compare and conclude

Since this value lands in a tail, we reject H_0 and accept H_1.

b) From table B for a two-tailed test at the $\alpha = 0.01$ level, the limit values are ± 2.861. Therefore, we once again reject H_0 and accept H_1.

Figure 11

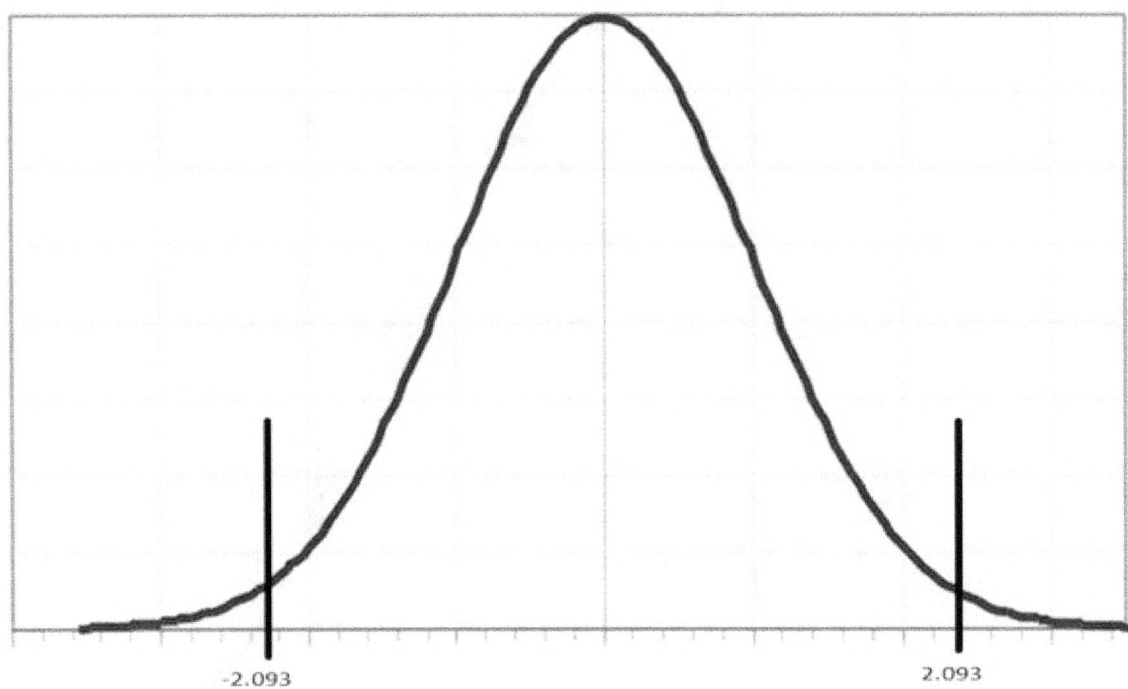

-2.093   2.093

Confidence intervals for the difference between two means, σ_1 and σ_2 known

21. a) Since both standard deviations are known, we use z scores, so for the 95% confidence interval we use 1.96

Difference of the two sample means is 0.1

$$\sqrt{\frac{\sigma_1^2}{n_1} + \frac{\sigma_2^2}{n_2}} = \sqrt{\frac{0.3^2}{25} + \frac{0.25^2}{36}} = \sqrt{.0036 + .0017} =$$

0.07.

Putting it all in the formula

$0.1 - 1.96 * 0.07 < \mu_1 - \mu_2 < 0.1 + 1.96 * 0.07 \rightarrow 0.1 - 0.14 < \mu_1 - \mu_2 < 0.1 + 0.14 \rightarrow$

$-0.04 < \mu_1 - \mu_2 < 0.24$.

b) For the 99% confidence interval we use z = 2.58. Putting it all into the formula

$0.1 - 2.58 * 0.07 < \mu_1 - \mu_2 < 0.1 + 2.58 * 0.07 \rightarrow 0.1 - 0.18 < \mu_1 - \mu_2 < 0.1 + 0.18 \rightarrow$

$-0.08 < \mu_1 - \mu_2 < 0.28$.

Note: The negative value means that $\mu_2 > \mu_1$.

22. Since the standard deviations are known, we will use z scores of 1.96 and 2.58.

a) Difference of the two sample means is -5

$$\sqrt{\frac{\sigma_1^2}{n_1} + \frac{\sigma_2^2}{n_2}} = \sqrt{\frac{2.5^2}{24} + \frac{2.1^2}{28}} = \sqrt{0.26 + 0.16} =$$

0.65.

Putting it all in the formula

$-5 - 1.96 * 0.65 < \mu_1 - \mu_2 < -5 + 1.96 * 0.65 \rightarrow -5 - 1.27 < \mu_1 - \mu_2 < -5 + 1.27 \rightarrow$

$-6.27 < \mu_1 - \mu_2 < -3.73$.

b) Using the z score of 2.58

$-5 - 2.58 * 0.65 < \mu_1 - \mu_2 < -5 + 2.58 * 0.65 \rightarrow -5 - 1.68 < \mu_1 - \mu_2 < -5 + 1.68 \rightarrow$

$-6.68 < \mu_1 - \mu_2 < -3.32$.

23. Since σ is known and is the same for both machines, we will use z scores.

a) For the 95% confidence interval, we use 1.96.

Difference of the two sample means is 0.2.

$$\sqrt{\frac{\sigma_1^2}{n_1} + \frac{\sigma_2^2}{n_2}} = \sqrt{\frac{0.85^2}{23} + \frac{0.85^2}{32}} = \sqrt{0.031 + 0.023} =$$

0.23.

Putting it all in the formula

$0.2 - 1.96 * 0.23 < \mu_1 - \mu_2 < 0.2 + 1.96 * 0.23 \rightarrow 0.2 - 0.45 < \mu_1 - \mu_2 < 0.2 + 0.45 \rightarrow$

$-0.25 < \mu_1 - \mu_2 < 0.65$.

b) For the 99% confidence interval we use 2.58.

Putting the numbers in the formula

$0.2 - 2.58 * 0.23 < \mu_1 - \mu_2 < 0.2 + 2.58 * 0.23 \rightarrow 0.2 - 0.59 < \mu_1 - \mu_2 < 0.2 + 0.59 \rightarrow$

$-0.39 < \mu_1 - \mu_2 < 0.79$.

Note: Once again, the negative number is a consequence of the fact that $\mu_2 > \mu_1$.

Confidence intervals for the difference between two means, σ_1 and σ_2 unknown

24. Since σ_1 and σ_2 are unknown we will use t scores with d. f. 18 - 1 = 17.

a) From table B we see that for a 95% confidence interval $t_{\alpha/2} = 2.110$. The difference between the two means is -5.

Difference of the two sample means is 0.2.

$$\sqrt{\frac{s_1^2}{n_1} + \frac{s_2^2}{n_2}} = \sqrt{\frac{3.7^2}{18} + \frac{4.1^2}{21}} = \sqrt{0.76 + 0.80} = 1.25.$$

Putting it all in the formula

$-5 - 2.110 * 1.25 < \mu_1 - \mu_2 < -5 + 2.110 * 1.25 \rightarrow -5 - 2.64 < \mu_1 - \mu_2 < -5 + 2.64 \rightarrow$

$-7.64 < \mu_1 - \mu_2 < -2.36.$

b) For the 99% confidence interval with d. f. = 17, $t_{\alpha/2} = 2.898$.

Putting it all in the formula

$-5 - 2.898 * 1.25 < \mu_1 - \mu_2 < -5 + 2.898 * 1.25 \rightarrow -5 - 3.62 < \mu_1 - \mu_2 < -5 + 3.62 \rightarrow$

$-8.62 < \mu_1 - \mu_2 < -1.38.$

25. Since σ_1 and σ_2 are unknown we will use t scores with d. f. 14 - 1 = 13.

a) From table B we see that for a 95% confidence interval $t_{\alpha/2} = 2.160$. The difference of the two sample means is 4.6.

$$\sqrt{\frac{s_1^2}{n_1} + \frac{s_2^2}{n_2}} = \sqrt{\frac{2.4^2}{14} + \frac{3.1^2}{16}} = \sqrt{0.41 + 0.60} = 1.01.$$

Putting it all in the formula

$4.6 - 2.160 * 1.01 < \mu_1 - \mu_2 < 4.6 + 2.160 * 1.01 \rightarrow 4.6 - 2.18 < \mu_1 - \mu_2 < 4.6 + 2.18 \rightarrow$

$2.42 < \mu_1 - \mu_2 < 6.78.$

b) For the 99% confidence interval with d. f. = 13, $t_{\alpha/2} = 3.012$.

Putting it all in the formula

$4.6 - 3.012 * 1.01 < \mu_1 - \mu_2 < 4.6 + 3.012 * 1.01 \rightarrow 4.6 - 3.04 < \mu_1 - \mu_2 < 4.6 + 3.04 \rightarrow$

$1.56 < \mu_1 - \mu_2 < 7.64.$

26. Since σ_1 and σ_2 are unknown, we will use t scores with d. f. = 11.

a) From table B, the $t_{\alpha/2}$ value for a 95% confidence interval is 2.201. The difference of the two sample means is -2.4.

$$\sqrt{\frac{s_1^2}{n_1} + \frac{s_2^2}{n_2}} = \sqrt{\frac{3.3^2}{12} + \frac{4.1^2}{14}} = \sqrt{0.91 + 1.20} = 1.45.$$

Putting it all in the formula

$-2.4 - 2.201 * 1.45 < \mu_1 - \mu_2 < -2.4 + 2.201 * 1.45 \rightarrow -2.4 - 3.19 < \mu_1 - \mu_2 < -2.4 + 3.19 \rightarrow$

$-5.59 < \mu_1 - \mu_2 < 0.79.$

b) For the 99% confidence interval with d. f. = 11, $t_{\alpha/2} = 3.106$.

Putting it all in the formula

$-2.4 - 3.106 * 1.45 < \mu_1 - \mu_2 < -2.4 + 3.106 * 1.45 \rightarrow -2.4 - 4.50 < \mu_1 - \mu_2 < -2.4 + 4.50 \rightarrow$

$-6.9 < \mu_1 - \mu_2 < 2.1.$

Testing the difference between two means, σ_1 and σ_2 known

27. Since both σ's are known, we will use z scores.

a)

Step 1: Establish H_0 and H_1. H_0: $\mu_1 = \mu_2$ or $\mu_1 - \mu_2 = 0$. H_1: $\mu_1 \neq \mu_2$ or $\mu_1 - \mu_2 \neq 0$.

Step 2: Draw the diagram and set the limits. This is seen in figure 12.

Step 3: Compute the test value

$$\sqrt{\frac{\sigma_1^2}{n_1} + \frac{\sigma_2^2}{n_2}} = \sqrt{\frac{0.4^2}{36} + \frac{0.4^2}{36}} = \sqrt{0.004 + 0.004} = 0.09.$$

Putting it all in the formula

$$z = \frac{16.1 - 15.9}{0.09} = 2.22.$$

Figure 12

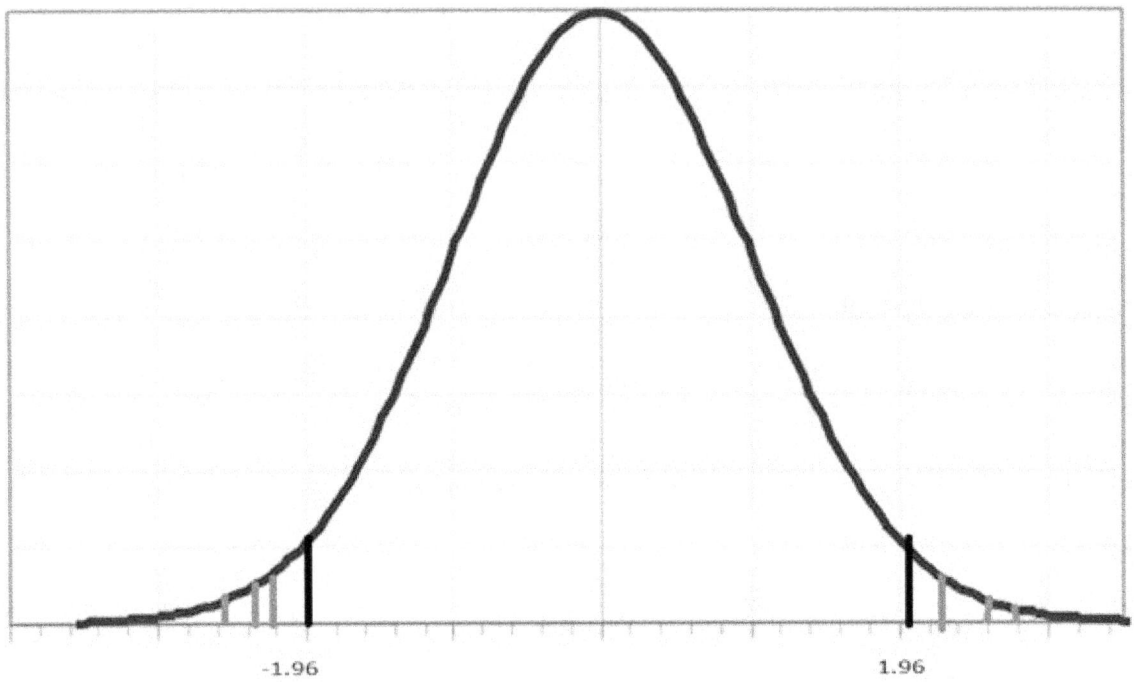

-1.96   1.96

Step 4: Compare and conclude.

Since 2.22 is greater than 1.96, we reject H_0 and accept H_1.

b) In this case the limits on the tails are ±2.58 and all other values are unchanged. Therefore, since 2.22 is between $\pm$ 2.58 we do not reject H_0.

28. Since both standard deviations are known, we use z scores. We will label the data from the businesses sample 1 and the data from the private residences sample 2.

a) Step 1: Establish H_0 and H_1. H_0: $\mu_1 = \mu_2$ or $\mu_1 - \mu_2 = 0$. H_1: $\mu_1 > \mu_2$ or $\mu_1 - \mu_2 > 0$.

Step 2: Draw the diagram and establish the limit(s). From the alternative hypothesis we know that this is a one-tailed test on the right. This is done in figure 13.

Step 3: Compute the test value

$$\sqrt{\dfrac{\sigma_1^2}{n_1} + \dfrac{\sigma_2^2}{n_2}} = \sqrt{\dfrac{0.18^2}{36} + \dfrac{0.12^2}{25}} = \sqrt{0.0009 + 0.0006} = 0.04.$$

Figure 13

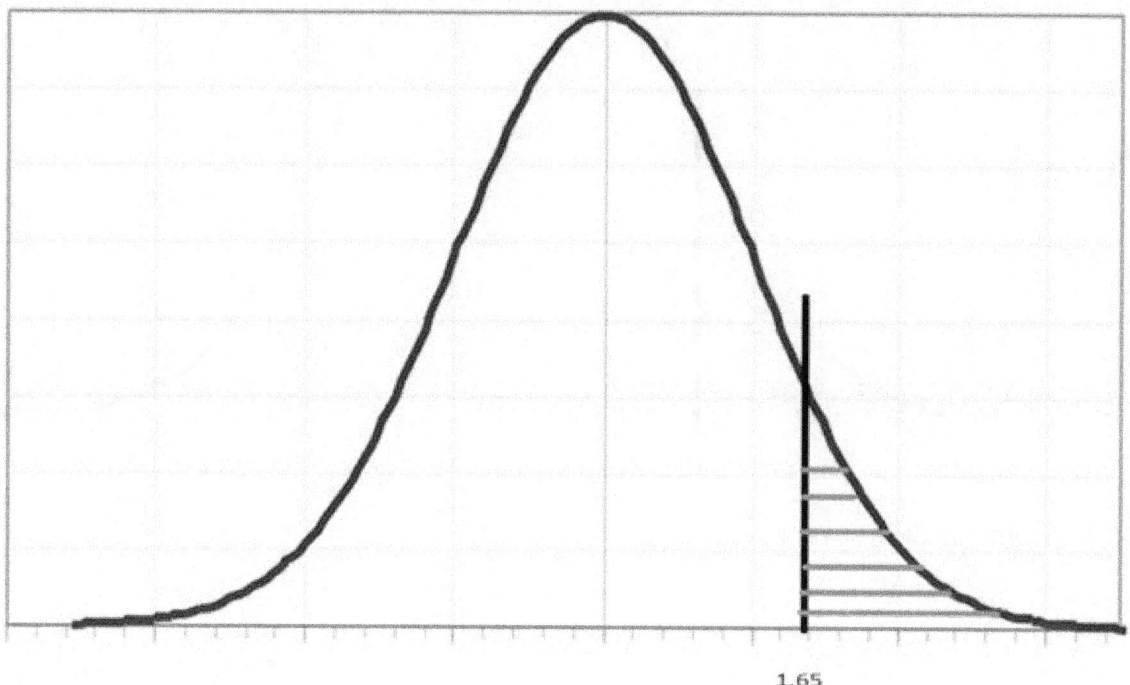

1.65

Putting it all in the formula

$$z = \dfrac{3.25 - 3.11}{0.04} = 3.5.$$

Step 4: Compare and conclude.

Since 3.5 is greater than 1.65 we reject H_0 and accept H_1.

b) For $\alpha = 0.01$ the limit value is 2.33 and all other values are the same. Therefore, we again reject H_0 and accept H_1.

29. Since both standard deviations are known, we will use z scores.

a) Step 1: Establish H_0 and H_1. Since the belief is that the average score on the second test is higher, this will be a one-tailed test with the tail on the left.

H_0: $\mu_1 = \mu_2$, H_1: $\mu_1 < \mu_2$.

Step 2: Draw the diagram and establish the limit(s). This is demonstrated in figure 14.

Figure 14

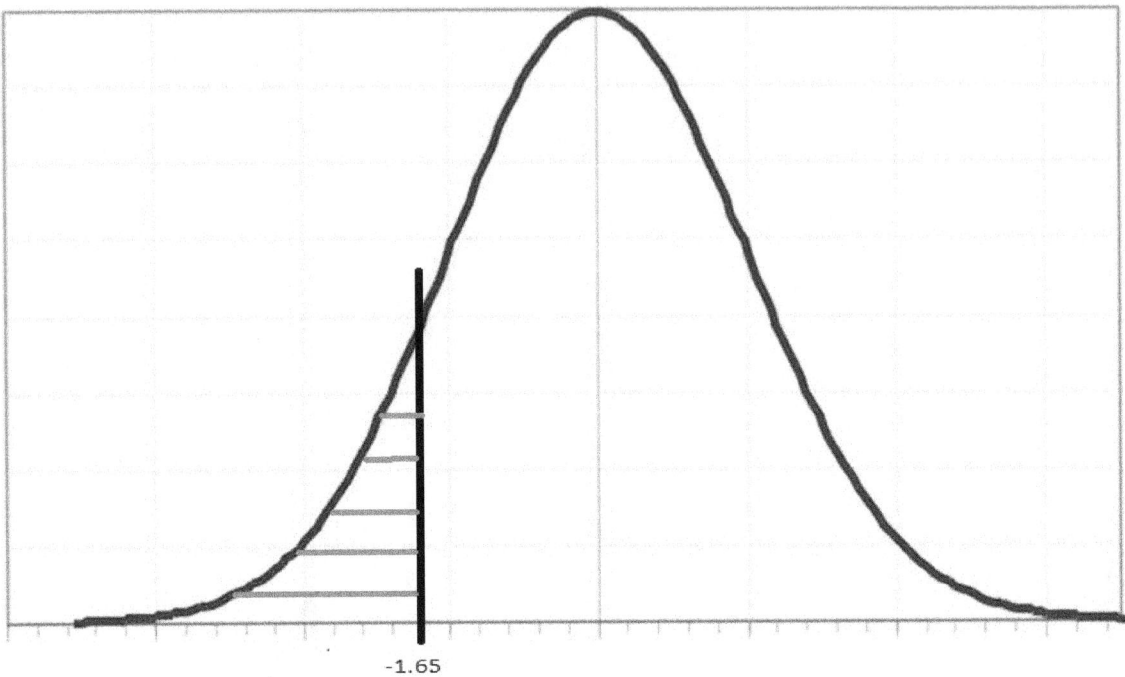

-1.65

Step 3: : Compute the test value

$$\sqrt{\frac{\sigma_1^2}{n_1} + \frac{\sigma_2^2}{n_2}} = \sqrt{\frac{2.65^2}{28} + \frac{3.15^2}{30}} = \sqrt{0.25 + 0.33} = 0.76.$$

Putting it all in the formula

$$z = \frac{92.4 - 94.2}{0.76} = -2.37.$$

Step 4: Compare and conclude. Since the test value is less than -1.65 we reject H_0 and accept H_1 at the $\alpha = 0.05$ level.

b) Since -2.37 is also less than -2.33, we also reject H_0 and accept H_1 at the $\alpha = 0.01$ level.

Testing the difference between two means, σ_1 and σ_2 unknown

30. Since we do not know σ_1 and σ_2, we must use t scores. The smallest sample size is 20, so d.f. = 19

a) Step 1: Establish the H_0 and H_1 hypotheses pairs.

32

$H_0: \mu_1 = \mu_2, H_1: \mu_1 \neq \mu_2.$

Step 2: Draw the diagram and establish the limits. This is a two-tailed test, so from table B, the boundary value of t is 2.093. This is illustrated in figure 15.

Figure 15

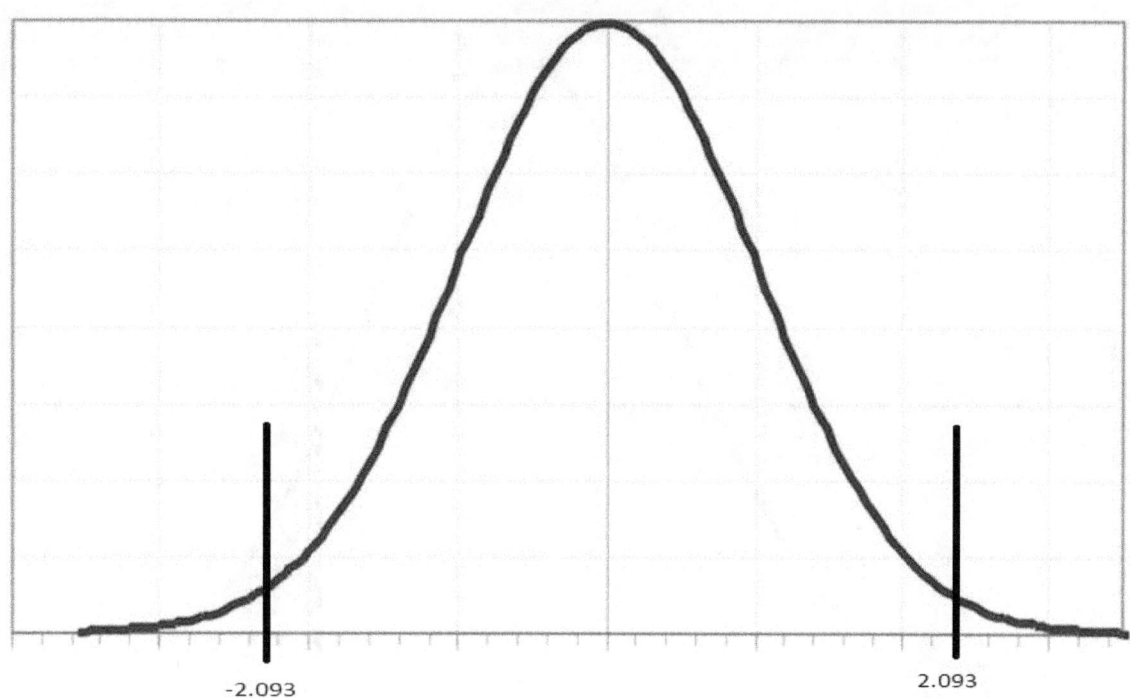

-2.093   2.093

Step 3: Compute the test value

$$\sqrt{\frac{s_1^2}{n_1} + \frac{s_2^2}{n_2}} = \sqrt{\frac{2.1^2}{20} + \frac{2.8^2}{28}} = \sqrt{0.22 + 0.28} = 0.71.$$

Putting it all in the formula

$$t = \frac{15.4 - 14.2}{0.71} = 1.69.$$

Step 4: Compare and conclude. Since this value is not in a tail, we cannot reject H_0.

b) Since we could not reject H_0 at the $\alpha = 0.05$ level we also cannot reject it at the $\alpha = 0.01$ level.

31. Since we do not know the values of σ_1 and σ_2 we will use t scores. The smallest sample size is 19, so d. f. is 18.

a) Step 1: Establish the H_0 and H_1 hypothesis pair. Since the belief is that the technique used on the first group is better, this is a one-tailed test on the right. H_0: $\mu_1 = \mu_2$, H_1: $\mu_1 > \mu_2$.

Step 2: Draw the diagram and establish the limit(s). This is a one-tailed test on the right, so from table B the value is 1.734. This is reflected in figure 16.

Figure 16

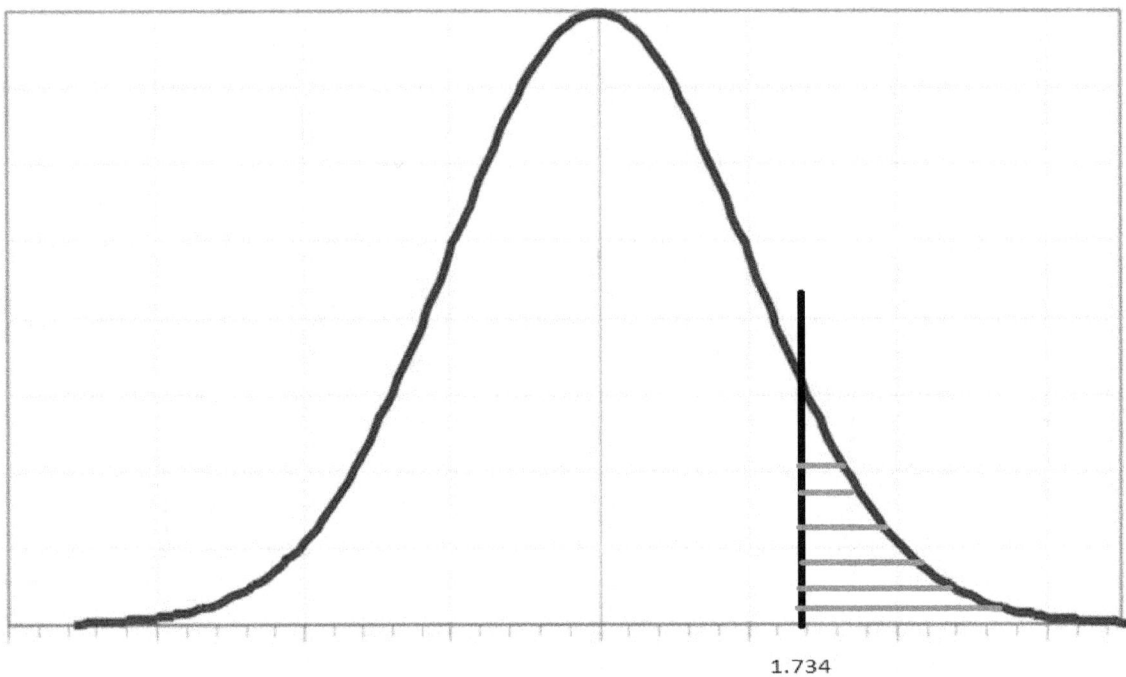

1.734

Step 3: Compute the test value

$$\sqrt{\frac{s_1^2}{n_1} + \frac{s_2^2}{n_2}} = \sqrt{\frac{3.6^2}{21} + \frac{2.9^2}{19}} = \sqrt{0.62 + 0.44} = 1.03.$$

Putting it all in the formula

$$t = \frac{86.5 - 80.2}{1.03} = 6.12.$$

Step 4: Compare and conclude. Since this value is in a tail, we reject H_0 and accept H_1.

b) From table B the value for $\alpha = 0.01$ is 2.552 so we also reject H_0 and accept H_1 at the $\alpha = 0.01$ level.

34

32. Since we do not know the values of σ_1 and σ_2 we will use t scores. The smallest sample size is 12, so d. f. is 11.

a) Step 1: Establish the H_0 and H_1 hypothesis pair. Since the belief is that the group by the chemical plant is having its growth stunted, this is a one-tailed test on the left. H_0: $\mu_1 = \mu_2$, H_1: $\mu_1 < \mu_2$.

Step 2: Draw the diagram and establish the limit(s). From table B, the value is 1.796, this can be seen in figure 17.

Figure 17

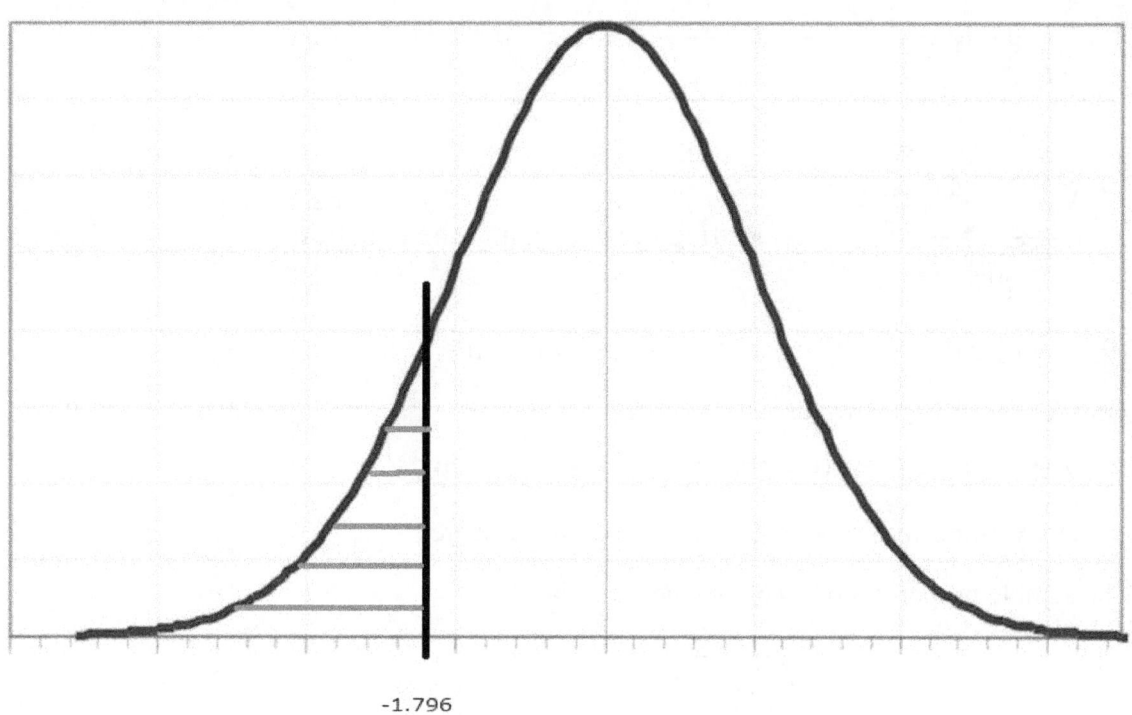

-1.796

Step 3: Compute the test value

$$\sqrt{\frac{s_1^2}{n_1} + \frac{s_2^2}{n_2}} = \sqrt{\frac{1.9^2}{12} + \frac{1.7^2}{13}} = \sqrt{0.30 + 0.22} = 0.72.$$

Putting it all in the formula

$$t = \frac{8.8 - 10.4}{0.72} = -2.22.$$

Step 4: Compare and conclude. Since the test value is in the tail we reject H_0 and accept H_1 at the $\alpha = 0.05$ level.

b) From table B the limit value is 2.718 so we do not reject H_0 at the $\alpha = 0.01$ level.

Confidence intervals for proportions

33. When working with proportions, we always use z scores.

a) The sample proportion $\acute{p} = 56 /300 = 0.19$, so the formula is

$$\acute{p} - 1.96 * \sqrt{\frac{\acute{p} * (1 - \acute{p})}{n}} < p < \acute{p} + 1.96 * \sqrt{\frac{\acute{p} * (1 - \acute{p})}{n}}$$

$$\sqrt{\frac{0.19 * 0.81}{300}} = 0.02.$$ Putting the values into the equation

$0.19 - 1.96 * 0.02 < p < 0.19 + 1.96 * 0.02 \rightarrow 0.15 < p < 0.23$.

b) For the $\alpha = 0.01$ level we use $z_{\alpha/2} = 2.58$.

$0.19 - 2.58 * 0.02 < p < 0.19 + 2.58 * 0.02 \rightarrow 0.14 < p < 0.24$.

34. When working with proportions, we always use z scores.

a) The sample proportion $\acute{p} = 0.67$, so the formula is

$$\acute{p} - 1.96 * \sqrt{\frac{\acute{p} * (1 - \acute{p})}{n}} < p < \acute{p} + 1.96 * \sqrt{\frac{\acute{p} * (1 - \acute{p})}{n}}$$

$$\sqrt{\frac{0.67 * 0.33}{400}} = 0.02.$$ Putting the values into the equation

$0.67 - 1.96 * 0.02 < p < 0.67 + 1.96 * 0.02 \rightarrow 0.63 < p < 0.71$.

b) For the $\alpha = 0.01$ level we use $z_{\alpha/2} = 2.58$.

$0.67 - 2.58 * 0.02 < p < 0.67 + 2.58 * 0.02 \rightarrow 0.62 < p < 0.72$.

35. Since it is a proportion, we will use z scores.

a) The sample proportion $\acute{p} = 0.20$, so the formula is

$$\acute{p} - 1.96 * \sqrt{\frac{\acute{p} * (1 - \acute{p})}{n}} < p < \acute{p} + 1.96 * \sqrt{\frac{\acute{p} * (1 - \acute{p})}{n}}$$

$$\sqrt{\frac{0.20 * 0.80}{500}} = 0.02.$$ Putting the values into the equation

$0.2 - 1.96 * 0.02 < p < 0.2 + 1.96 * 0.02 \rightarrow 0.16 < p < 0.24.$

b) For the $\alpha = 0.01$ level we use $z_{\alpha/2} = 2.58$.

$0.2 - 2.58 * 0.02 < p < 0.2 + 2.58 * 0.02 \rightarrow 0.15 < p < 0.25.$

Hypothesis testing with proportions

36. Since it is a proportion, we will use z scores.

a) Step 1: Establish H_0 and H_1. H_0: $p = 0.55$, H_1: $p \neq 0.55$

Step 2: Draw the diagram and establish the limit(s), the limit score in this case is 1.96, as can be seen in figure 18.

Step 3: Compute the test value. The denominator is

$$\sqrt{\frac{p * (1-p)}{n}} = \sqrt{\frac{0.55 * 0.45}{400}} = 0.025$$

$$\frac{\acute{p} - p}{0.025} = \frac{0.50 - 0.55}{0.025} = -2.$$

Step 4: Compare and conclude. Since the test value is in a tail, we reject H_0 and accept H_1.

b) For the $\alpha = 0.01$ level, the limits are ± 2.58, so we do NOT reject H_0.

Figure 18

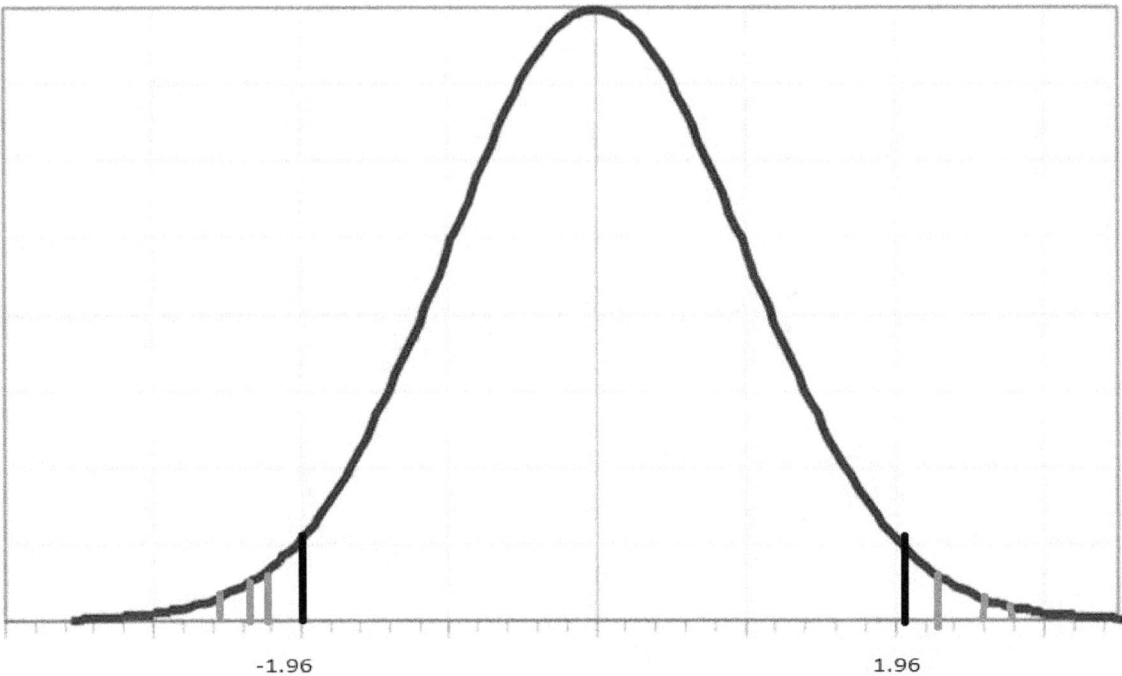

37. Since we are dealing with proportions, we will use z scores.

a) Step 1: Establish H_0 and H_1.

H_0: p = 0.12, H_1: p > 0.12.

Step 2: Draw the diagram and establish the limits, this will be a one-tailed test on the right. This is seen in figure 19.

Step 3: Compute the test value. The denominator is

$$\sqrt{\frac{p * (1-p)}{n}} = \sqrt{\frac{0.12 * 0.88}{1000}} = 0.01$$

$$\frac{\acute{p} - p}{0.01} = \frac{0.21 - 0.12}{0.01} = 9.$$

Step 4: Since the value is in the tail, we reject H_0 and accept H_1.

b) The limit for the one-tailed test at the α = 0.01 level is 2.33, so the value is in the tail and we reject H_0 and accept H_1.

Figure 19

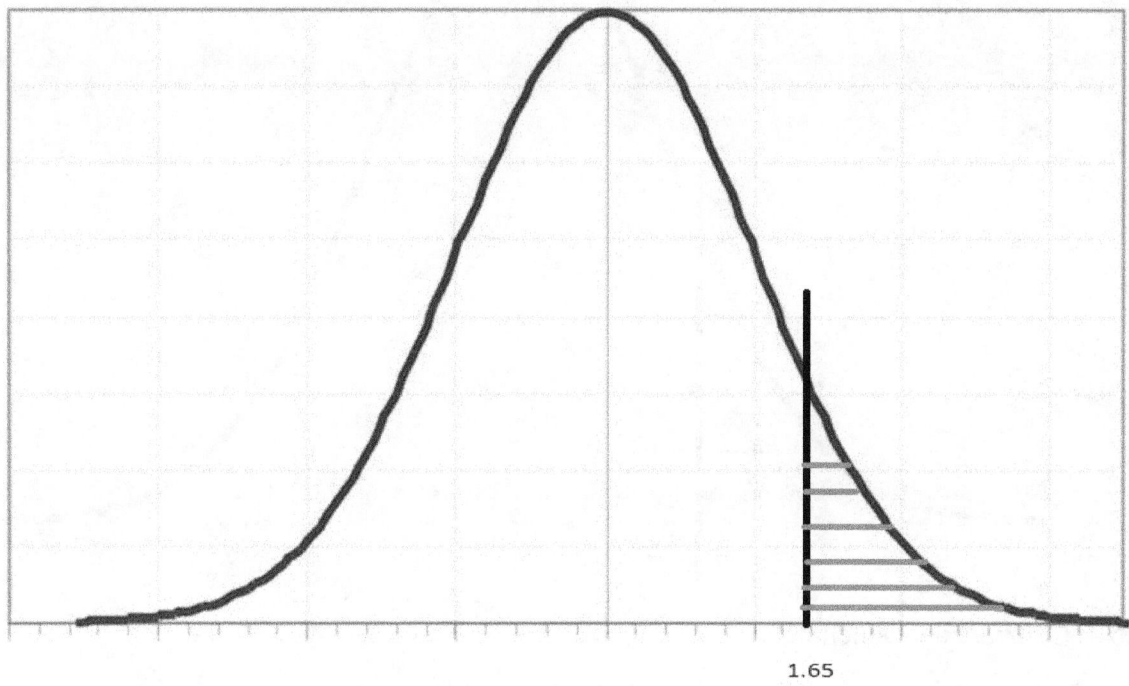

1.65

38. Since we are dealing with proportions, we will use z scores.

a) Step 1: Establish H_0 and H_1.

H_0: p = 0.55, H_1: p < 0.55.

Step 2: Draw the diagram and establish the limits, this will be a one-tailed test on the left. This is seen in figure 20.

Step 3: Compute the test value

The denominator is

$$\sqrt{\frac{p * (1-p)}{n}} = \sqrt{\frac{0.55 * 0.45}{660}} = 0.019$$

$$\frac{\acute{p} - p}{0.019} = \frac{0.48 - 0.55}{0.019} = -3.68.$$

39

Figure 20

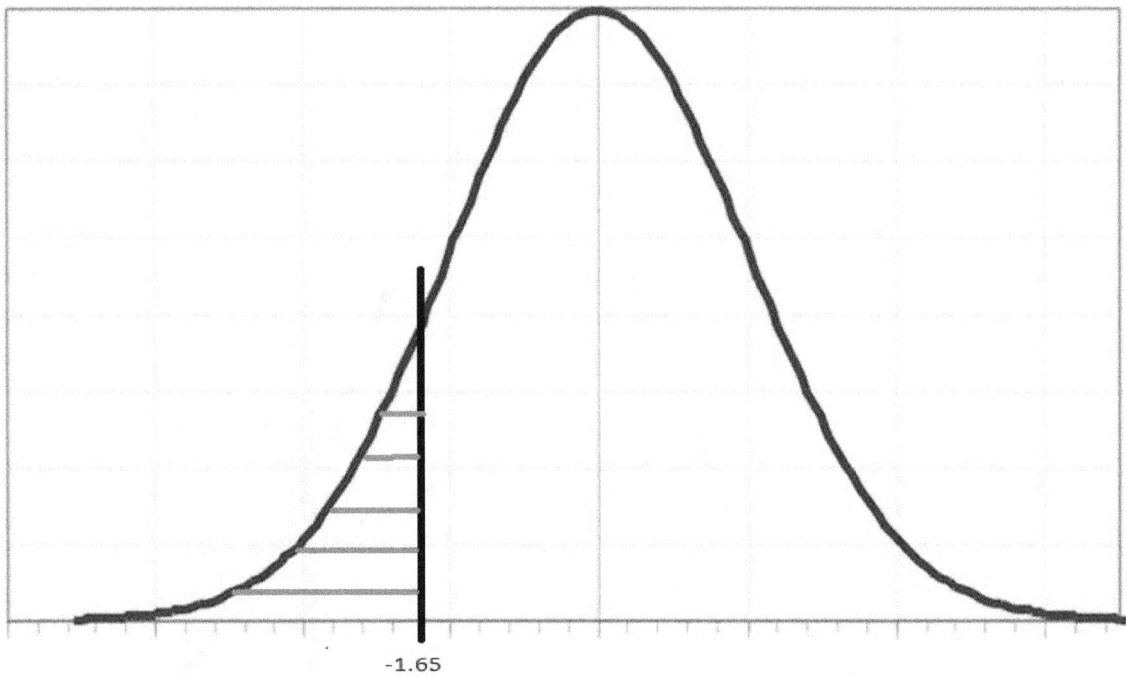

-1.65

Step 4: Compare and conclude

Since the test value is in the tail, we reject H_0 and accept H_1.

b) The limit of the tail for the $\alpha = 0.01$ level is -2.33 so once again we reject H_0 and accept H_1.

Hypothesis testing with correlation coefficients

39. A company claims that their method of preparation for the ACT exam is superior and there is a direct correlation between the hours of instruction and the increase in scores. Eight students are tested and the data is summarized in the following table

Expanding the table so that we can more easily compute the correlation coefficient

Hours of instruction x	x^2	xy	Score on test y	y^2
2	4	51.2	25.6	655.36
2.5	6.25	65	26	676
3.0	9	78.3	26.1	681.21
3.5	12.25	91	26	676
3.5	12.25	92.4	26.4	696.96
4.0	16	107.6	26.9	723.61
4.5	20.25	125.55	27.9	778.41
5.0	25	136.5	27.3	745.29
$\sum x = 28$	$\sum x^2 = 105$	$\sum xy = 747.55$	$\sum y = 212.2$	$\sum y^2 = 5632.84$

The computation of the correlation coefficient (r) is done using the following formula

$$r = \frac{n(\sum xy) - (\sum x)(\sum y)}{\sqrt{[\, n(\sum x^2) - (\sum x)^2\,]\,[\, n(\sum y^2) - (\sum y)^2\,]}}$$

where n is the number of data PAIRS.

The computation of the numerator is

$8 * 747.55 - 28 * 212.2 = 5980.4 - 5941.6 = 38.8$.

The computation in the denominator involving the x's is

$8 * 105 - 28^2 = 840 - 784 = 56$.

The computation in the denominator involving the y's is

$8 * 5632.84 - 212.2^2 = 45062.72 - 45028.84 = 33.88$.

$56 * 33.88 = 1897.28$ and the square root of this is 43.56. The value of r is then

$r = 38.8 / 43.56 = 0.89$

Step 1: Establish H_0 and H_1. H_0: $\rho = 0$, H_1: $\rho > 0$.

Step 2: Draw the diagram and set the limit. Since n = 8, d.f. = 8 – 2 or 6. It will be a one-tailed test on the right using t scores. From table B the value is 1.943 and this is seen in figure 21

Figure 21

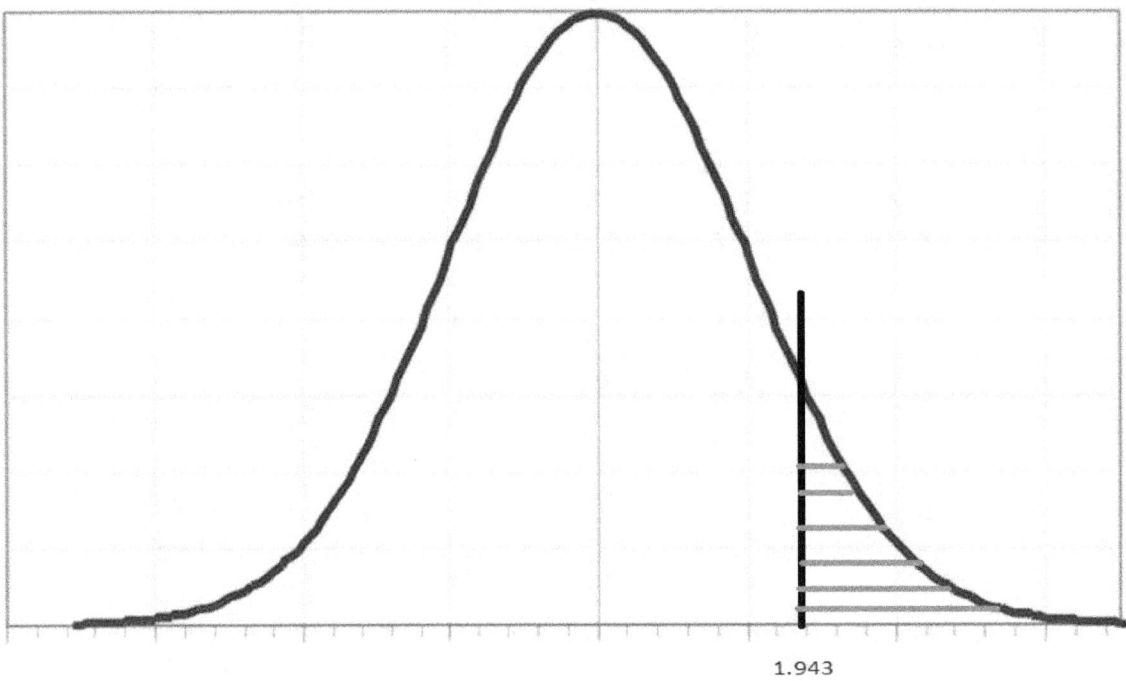

1.943

Step 3: Compute the test value

$$t = r \sqrt{\dfrac{n-2}{1-r^2}} \qquad t = 0.89 \sqrt{\dfrac{6}{1-0.89^2}} \qquad t = 4.76.$$

Step 4: Compare and conclude. Since the computed value is greater than the limit, we reject H_0 and accept H_1.

b) The formulas for a and b of the regression line are

$$a = \frac{(\sum y)(\sum x^2) - (\sum x)(\sum xy)}{n(\sum x^2) - (\sum x)^2}$$

$$b = \frac{n(\sum xy) - (\sum x)(\sum y)}{n(\sum x^2) - (\sum x)^2}.$$

The computation of the denominator of both values is

42

$8 * 105 - 28^2 = 840 - 784 = 56$.

The numerator of a is

$212.2 * 105 - 28 * 747.55 = 22281 - 20931.4 = 1349.6$, making the value of

$a = 1349.6 / 56 = 24.1$.

The numerator of b is

$8 * 747.55 - 28 * 212.2 = 5980.4 - 5941.6 = 38.8$, making the value of

$b = 38.8 / 56 = 0.69$. The equation of the regression line is

$\hat{y} = 24.1 + 0.69 * x$.

c) $\hat{y} = 24.1 + 0.69 * 6 = 28.24$.

40. The first step is to expand the table so that it can hold the computed values

Average number of text messages x	x^2	xy	Score on exam y	y^2
4.4	19.36	396	90	8100
3.6	12.96	331.2	92	8464
5.6	31.36	492.8	88	7744
5.2	27.04	457.6	88	7744
6.9	47.61	593.4	86	7396
7.8	60.84	655.2	84	7056
7.2	51.84	590.4	82	6724
8.4	70.56	697.2	83	6889
$\sum x = 49.1$	$\sum x^2 = 321.57$	$\sum xy = 4213.8$	$\sum y = 693$	$\sum y^2 = 60117$

a) The computation of the correlation coefficient (r) is done using the following formula

$$r = \frac{n(\sum xy) - (\sum x)(\sum y)}{\sqrt{[\ n(\sum x^2) - (\sum x)^2\ ][\ n(\sum y^2) - (\sum y)^2\ ]}}$$

where n is the number of data PAIRS.

The computation of the numerator is

  8 * 4213.8 – 49.1 * 693 = 33710.4 – 34026.3 = -315.9.

The computation in the denominator involving the x's is

8 * 321.57 – 49.1^2 = 2572.56 – 2410.81 = 161.75.

The computation in the denominator involving the y's is

8 * 60117 – 693^2 = 480936 – 480249 = 687.

161.75 * 687 = 111122.25 and the square root of this is 333.35. The value of r is then

r = -315.9 / 333.35 = -0.95.

Step 1: Establish H_0 and H_1. H_0: $\rho = 0$, H_1: $\rho < 0$.

Step 2: Draw the diagram and set the limit(s). With 8 pairs the d. f. is 6 and this is a one-tailed test on the left. From table B, the value is -1.943 and this is summarized in figure 22.

Figure 22

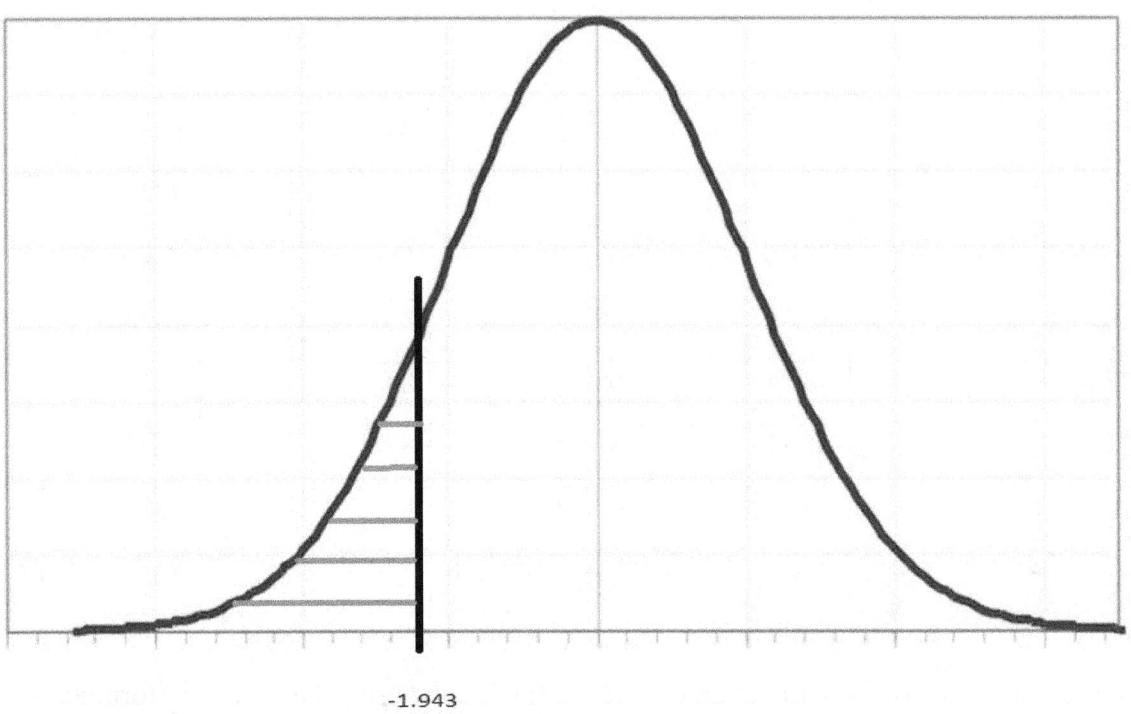

-1.943

Step 3: Compute the test value

44

Compute the test value

$$t = r\sqrt{\frac{n-2}{1-r^2}} \qquad t = -0.95\sqrt{\frac{6}{1-(-0.95)^2}} \qquad t = -7.36.$$

Step 4: Compare and conclude. Since the test value is in the tail we reject H_0 and accept H_1.

b) The formulas for a and b of the regression line are

$$a = \frac{(\sum y)(\sum x^2) - (\sum x)(\sum xy)}{n(\sum x^2) - (\sum x)^2}$$

$$b = \frac{n(\sum xy) - (\sum x)(\sum y)}{n(\sum x^2) - (\sum x)^2}.$$

The computation of the denominator of both values is

$8 * 321.57 - (49.1)^2 = 2572.56 - 2410.81 = 161.75.$

The numerator of a is

$693 * 321.57 - 49.1 * 4213.8 = 222848.01 - 206897.58 = 15950.43$

making the value of a

$a = 15950.43 / 161.75 = 98.61.$

The computation of the numerator of b is

$8 * 4213.8 - 49.1 * 693 = 33710.4 - 34026.3 = -315.9$

making the value of b

$b = -315.9 / 161.75 = -1.95.$

The equation for the linear regression line is then

$\hat{y} = 98.61 - 1.95 * x.$

c) Putting 10 in for x, the value is

$\hat{y} = 98.61 - 1.95 * 10 = 79.11.$

41. The first step is to fill out the table

Hours per week playing video games x	x^2	xy	Weight y	y^2
2.5	6.25	196	78.4	6146.56
2.2	4.84	169.4	77	5929
2.0	4	146	73	5329
2.8	7.84	221.2	79	6241
3.4	11.56	289	85	7225
3.1	9.61	269.7	87	7569
3.0	9	258	86	7396
2.9	8.41	229.1	79	6241
$\sum x = 21.9$	$\sum x^2 = 61.51$	$\sum xy = 1778.4$	$\sum y = 644.4$	$\sum y^2 = 52076.56$

a) The computation of the correlation coefficient (r) is done using the following formula

$$r = \frac{n(\sum xy) - (\sum x)(\sum y)}{\sqrt{[\, n(\sum x^2) - (\sum x)^2 \,][\, n(\sum y^2) - (\sum y)^2 \,]}}$$

where n is the number of data PAIRS.

The computation of the numerator is

$8 * 1778.4 - 21.9 * 644.4 = 14227.2 - 14112.36 = 114.84$.

The computation in the denominator involving the x's is

$8 * 61.51 - 21.9^2 = 492.08 - 479.61 = 12.47$.

The computation in the denominator involving the y's is

$8 * 52076.56 - 644.4^2 = 416612.48 - 415251.36 = 1361.12$.

$12.47 * 1361.12 = 16973.166$ which has the square root 130.28. Therefore, the value of r is

$r = 114.84 / 130.28 = 0.88$.

Step 1: Determine H_0 and H_1 hypotheses. H_0: $\rho = 0$, H_1: $\rho > 0$.

Step 2: Draw the diagram and establish the limits. The d. f. is the number of data pairs minus 2 or $8 - 2 = 6$. From table B, we can see that the limit value is for $\alpha = 0.01$ is 3.143. This is illustrated in figure 23.

Figure 23

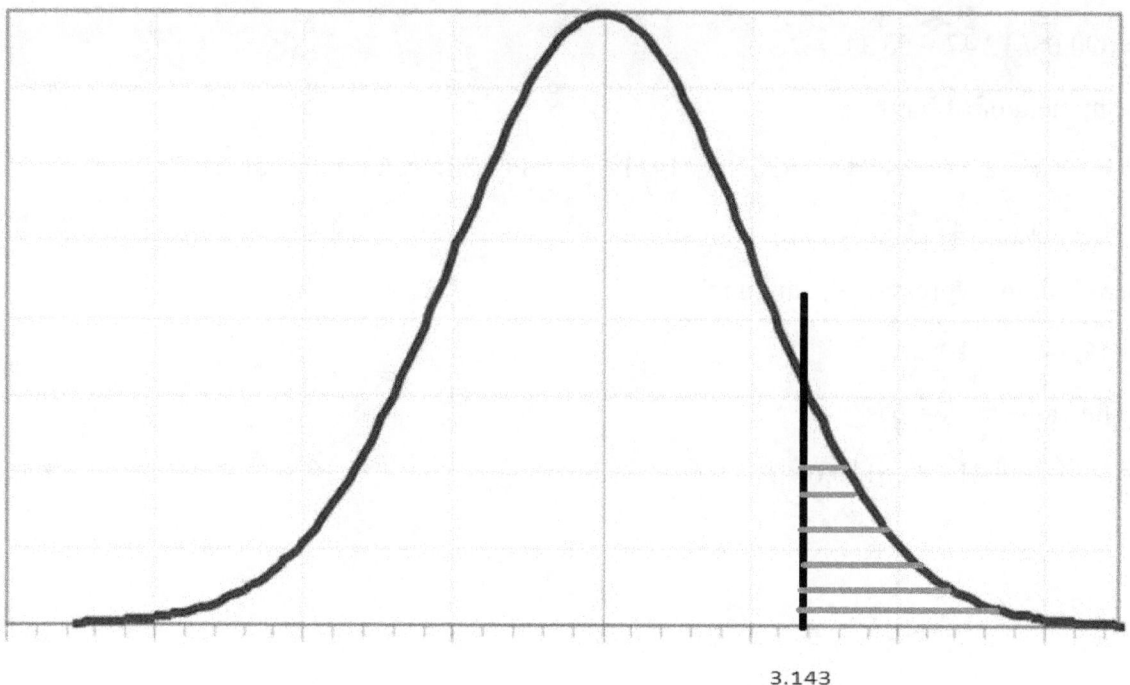

3.143

Step 3: Compute the test value

Compute the test value

$$t = 0.88 * \sqrt{\frac{6}{1 - (0.88)^2}} \quad \rightarrow \quad t = 4.49.$$

Step 4: Compare and conclude. Since the test value is in the tail we reject H_0 and accept H_1.

b) The formulas for a and b of the regression line are

$$a = \frac{(\sum y)(\sum x^2) - (\sum x)(\sum xy)}{n(\sum x^2) - (\sum x)^2}$$

$$b = \frac{n(\sum xy) - (\sum x)(\sum y)}{n(\sum x^2) - (\sum x)^2} \, .$$

The computation of the denominator of both values is

$8 * 61.51 - 21.9^2 = 492.08 - 479.61 = 12.47$.

The numerator of a is

$644.4 * 61.51 - 21.9 * 1778.4 = 39637.04 - 38946.96 = 690.08$, making the value of a

$a = 690.08/ 12.47 = 55.34$.

The numerator of b is

$8 * 1778.4 - 21.9 * 644.4 = 14227.2 - 14112.36 = 114.84$, making the value of b

$b = 114.84 / 12.47 = 9.21$.

Therefore, the regression equation is

$\hat{y} = 55.34 + 9.21 * x$.

c) The value for $x = 6$ is

$\hat{y} = 55.34 + 9.21 * 6 = 110.6$.

Table A: The Standard Normal Distribution
Cumulative Values (Area to the left)

z	.00	.01	.02	.03	.04	.05	.06	.07	.08	.09
-3.4	.0003	.0003	.0003	.0003	.0003	.0003	.0003	.0003	.0003	.0002
-3.3	.0005	.0005	.0005	.0004	.0004	.0004	.0004	.0004	.0004	.0003
-3.2	.0007	.0007	.0006	.0006	.0006	.0006	.0006	.0005	.0005	.0005
-3.1	.0010	.0009	.0009	.0009	.0008	.0008	.0008	.0008	.0007	.0007
-3.0	.0013	.0013	.0013	.0012	.0012	.0011	.0011	.0011	.0010	.0010
-2.9	.0019	.0018	.0018	.0017	.0016	.0016	.0015	.0015	.0014	.0014
-2.8	.0026	.0025	.0024	.0023	.0023	.0022	.0021	.0021	.0020	.0019
-2.7	.0035	.0034	.0033	.0032	.0031	.0030	.0029	.0028	.0027	.0026
-2.6	.0047	.0045	.0044	.0043	.0041	.0040	.0039	.0038	.0037	.0036
-2.5	.0062	.0060	.0059	.0057	.0055	.0054	.0052	.0051	.0049	.0048
-2.4	.0082	.0080	.0078	.0075	.0073	.0071	.0069	.0068	.0066	.0064
-2.3	.0107	.0104	.0102	.0099	.0096	.0094	.0091	.0089	.0087	.0084
-2.2	.0139	.0136	.0132	.0129	.0125	.0122	.0119	.0116	.0113	.0110
-2.1	.0179	.0174	.0170	.0166	.0162	.0158	.0154	.0150	.0146	.0143
-2.0	.0228	.0222	.0217	.0212	.0207	.0202	.0197	.0192	.0188	.0183
-1.9	.0287	.0281	.0274	.0268	.0262	.0256	.0250	.0244	.0239	.0233
-1.8	.0359	.0351	.0344	.0336	.0329	.0322	.0314	.0307	.0301	.0294
-1.7	.0446	.0436	.0427	.0418	.0409	.0401	.0392	.0384	.0375	.0367
-1.6	.0548	.0537	.0526	.0516	.0505	.0495	.0485	.0475	.0465	.0455
-1.5	.0668	.0655	.0643	.0630	.0618	.0606	.0594	.0582	.0571	.0559
-1.4	.0808	.0793	.0778	.0764	.0749	.0735	.0721	.0708	.0694	.0681
-1.3	.0968	.0951	.0934	.0918	.0901	.0885	.0869	.0853	.0838	.0823
-1.2	.1151	.1131	.1112	.1093	.1075	.1056	.1038	.1020	.1003	.0985
-1.1	.1357	.1335	.1314	.1292	.1271	.1251	.1230	.1210	.1190	.1170
-1.0	.1587	.1562	.1539	.1515	.1492	.1469	.1446	.1423	.1401	.1379
-0.9	.1841	.1814	.1788	.1762	.1736	.1711	.1685	.1660	.1635	.1611
-0.8	.2119	.2090	.2061	.2033	.2005	.1977	.1949	.1922	.1894	.1867
-0.7	.2420	.2389	.2358	.2327	.2296	.2266	.2236	.2206	.2177	.2148
-0.6	.2743	.2709	.2676	.2643	.2611	.2578	.2546	.2514	.2483	.2451
-0.5	.3085	.3050	.3015	.2981	.2946	.2912	.2877	.2843	.2810	.2776
-0.4	.3446	.3409	.3372	.3336	.3300	.3264	.3228	.3192	.3156	.3121
-0.3	.3821	.3783	.3745	.3707	.3669	.3632	.3594	.3557	.3520	.3483
-0.2	.4207	.4168	.4129	.4090	.4052	.4013	.3974	.3936	.3897	.3859
-0.1	.4602	.4562	.4522	.4483	.4443	.4404	.4364	.4325	.4286	.4247
-0.0	.5000	.4960	.4920	.4880	.4840	.4801	.4761	.4721	.4681	.4641

For all values less than -3.49, use .0001.

z	.00	.01	.02	.03	.04	.05	.06	.07	.08	0.09
0.0	.5000	.5040	.5080	.5120	.5160	.5199	.5239	.5279	.5319	.5359
0.1	.5398	.5438	.5478	.5517	.5557	.5596	.5636	.5675	.5714	.5753
0.2	.5793	.5832	.5871	.5910	.5948	.5987	.6026	.6064	.6103	.6141
0.3	.6179	.6217	.6255	.6293	.6331	.6368	.6406	.6443	.6480	.6517
0.4	.6554	.6591	.6628	.6664	.6700	.6736	.6772	.6808	.6844	.6879
0.5	.6915	.6950	.6985	.7019	.7054	.7088	.7123	.7157	.7190	.7224
0.6	.7257	.7291	.7324	.7357	.7389	.7422	.7454	.7486	.7517	.7549
0.7	.7580	.7611	.7642	.7673	.7704	.7734	.7764	.7794	.7823	.7852
0.8	.7881	.7910	.7939	.7967	.7995	.8023	.8051	.8078	.8106	.8133
0.9	.8159	.8186	.8212	.8238	.8264	.8289	.8315	.8340	.8365	.8389
1.0	.8413	.8438	.8461	.8485	.8508	.8531	.8554	.8577	.8599	.8621
1.1	.8643	.8665	.8686	.8708	.8729	.8749	.8770	.8790	.8810	.8830
1.2	.8849	.8869	.8888	.8907	.8925	.8944	.8962	.8980	.8997	.9015
1.3	.9032	.9049	.9066	.9082	.9099	.9115	.9131	.9147	.9162	.9177
1.4	.9192	.9207	.9222	.9236	.9251	.9265	.9279	.9292	.9306	.9319
1.5	.9332	.9345	.9357	.9370	.9382	.9394	.9406	.9418	.9429	.9441
1.6	.9452	.9463	.9474	.9484	.9495	.9505	.9515	.9525	.9535	.9545
1.7	.9554	.9564	.9573	.9582	.9591	.9599	.9608	.9616	.9625	.9633
1.8	.9641	.9649	.9656	.9664	.9671	.9678	.9686	.9693	.9699	.9706
1.9	.9713	.9719	.9726	.9732	.9738	.9744	.9750	.9756	.9761	.9767
2.0	.9772	.9778	.9783	.9788	.9793	.9798	.9803	.9808	.9812	.9817
2.1	.9821	.9826	.9830	.9834	.9838	.9842	.9846	.9850	.9854	.9857
2.2	.9861	.9864	.9868	.9871	.9875	.9878	.9881	.9884	.9887	.9890
2.3	.9893	.9896	.9898	.9901	.9904	.9906	.9909	.9911	.9913	.9916
2.4	.9918	.9920	.9922	.9925	.9927	.9929	.9931	.9932	.9934	.9936
2.5	.9938	.9940	.9941	.9943	.9945	.9946	.9948	.9949	.9951	.9952
2.6	.9953	.9955	.9956	.9957	.9959	.9960	.9961	.9962	.9963	.9964
2.7	.9965	.9966	.9967	.9968	.9969	.9970	.9971	.9972	.9973	.9974
2.8	.9974	.9975	.9976	.9977	.9977	.9978	.9979	.9979	.9980	.9981
2.9	.9981	.9982	.9982	.9983	.9984	.9984	.9985	.9985	.9986	.9986
3.0	.9987	.9987	.9987	.9988	.9988	.9989	.9989	.9989	.9990	.9990
3.1	.9990	.9991	.9991	.9991	.9992	.9992	.9992	.9992	.9993	.9993
3.2	.9993	.9993	.9994	.9994	.9994	.9994	.9994	.9995	.9995	.9995
3.3	.9995	.9995	.9995	.9996	.9996	.9996	.9996	.9996	.9996	.9997
3.4	.9997	.9997	.9997	.9997	.9997	.9997	.9997	.9997	.9997	.9998

For all values greater than 3.49, use 0.9999.

Table B: Values of the t Distribution

	Confidence intervals	90%	95%	98%	99%
	One tail α	0.05	0.025	0.01	0.005
d. f.	Two tail α	0.10	0.05	0.02	0.01
1		6.314	12.706	31.821	63.657
2		2.920	4.303	6.965	9.925
3		2.353	3.182	4.541	5.841
4		2.132	2.776	3.747	4.604
5		2.015	2.571	3.365	4.032
6		1.943	2.447	3.143	3.707
7		1.895	2.365	2.998	3.499
8		1.860	2.306	2.896	3.355
9		1.833	2.262	2.821	3.250
10		1.812	2.228	2.764	3.169
11		1.796	2.201	2.718	3.106
12		1.782	2.179	2.681	3.055
13		1.771	2.160	2.650	3.012
14		1.761	2.145	2.624	2.977
15		1.753	2.131	2.602	2.947
16		1.746	2.120	2.583	2.921

17		1.740	2.110	2.567	2.989
18		1.734	2.101	2.552	2.878
19		1.729	2.093	2.539	2.861
20		1.725	2.086	2.528	2.845
21		1.721	2.080	2.518	2.831
22		1.717	2.074	2.508	2.819
23		1.714	2.069	2.500	2.807
24		1.711	2.064	2.492	2.797
25		1.708	2.060	2.485	2.787
26		1.706	2.056	2.479	2.779
27		1.703	2.052	2.473	2.771
28		1.701	2.048	2.467	2.763
29		1.699	2.045	2.462	2.756
30		1.697	2.042	2.457	2.750
40		1.684	2.021	2.423	2.704
60		1.671	2.000	2.390	2.660
z-values		1.645	1.960	2.326	2.576

BOOKS IN RECREATIONAL MATHEMATICS BY CHARLES ASHBACHER AND ASSOCIATES

Topics in Recreational Mathematics 1/2015 ISBN 978-1507603215

Topics in Recreational Mathematics 2/2015 ISBN 978-1508617099

Topics in Recreational Mathematics 3/2015 ISBN 978-1511641005

Topics in Recreational Mathematics 4/2015 ISBN 978-1514317518

Alphametics as Expressed in Recreational Mathematics Magazine ISBN 978-1508538134

Ten Year Cumulative Index to the Journal of Recreational Mathematics, edited by Joseph S. Madachy and Charles Ashbacher ISBN 978-1508936800

Alphametics Expressing Thoughts From the Star Trek Original Series ISBN 978-1512152784

Mathematical Cartoons ISBN 978-1514207130

Associates

Artist Catie Ribble

Editor Rachel Pollari

Editor Jennifer Corrigan

Artist Jenna Richardson

www.ingramcontent.com/pod-product-compliance
Lightning Source LLC
Chambersburg PA
CBHW080612180526
45168CB00007B/2882